Electrical
Discipline-Specific Review for the FE/EIT Exam

Second Edition

Robert B. Angus, PE
John E. Hajjar
Abdulrahman Yassine
with Michael R. Lindeburg, PE

Professional Publications, Inc. • Belmont, California

Benefit by Registering This Book with PPI

- Get book updates and corrections
- Hear the latest exam news
- Obtain exclusive exam tips and strategies
- Receive special discounts

Register your book at **www.ppi2pass.com/register**.

Report Errors and View Corrections for This Book

PPI is grateful to every reader who notifies us of a possible error. Your feedback allows us to improve the quality and accuracy of our products. You can report errata and view corrections at **www.ppi2pass.com/errata**.

ELECTRICAL DISCIPLINE-SPECIFIC REVIEW FOR THE FE/EIT EXAM

Current printing of this edition: 5

Printing History

edition number	printing number	update
2	3	Minor corrections.
2	4	Minor corrections. Copyright update.
2	5	Minor corrections.

Copyright © 2010 by Professional Publications, Inc. (PPI). All rights reserved. No part of this publication may be reproduced, stored in a retrieval system, or transmitted, in any form or by any means, electronic, mechanical, photocopying, recording, or otherwise, without the prior written permission of the publisher.

Printed in the United States of America.

PPI
1250 Fifth Avenue, Belmont, CA 94002
(650) 593-9119
www.ppi2pass.com

Library of Congress Cataloging-in-Publication Data
Electrical discipline-specific review for the FE/EIT exam / Robert B. Angus ... [et al.].--
    2nd ed.
       p.   cm.
   ISBN: 978-1-59126-066-0
   1. Engineering--United States--Examinations--Study guides. 2. Electric engineering--United States--Examinations--Study guides 3. Engineering--Problems, exercises, etc. 4. Engineers--Certification--United States. I. Angus, Robert B. (Robert Brownell)

TA159.A2 2006
621.3076--dc22

2005057892

Table of Contents

Preface and Acknowledgments v

Engineering Registration in the United States vii

Common Questions About the DS Exam xiii

How to Use This Book xv

Nomenclature 1

Practice Problems
Circuits 5
Power 8
Electromagnetics 11
Control Systems 12
Communications 14
Signal Processing 15
Electronics 17
Digital Systems 19
Computer Systems 20

Practice Exam 1
Problems 23
Solutions 30

Practice Exam 2
Problems 39
Solutions 49

Preface and Acknowledgments

This book is one in a series intended for engineers and students who are taking a discipline-specific (DS) afternoon session of the Fundamentals of Engineering (FE) exam.

The topics covered in the DS afternoon FE exams are completely different from the topics covered in the morning session of the FE exam. Since this book only covers one discipline-specific exam, it provides exam-level problems that are like those found on the afternoon half of the FE exam for the Electrical discipline.

This book is intended to be a quick review of the material unique to the afternoon session of the Electrical engineering exam. The material presented covers the subjects most likely to be on the exam. This book is not a thorough treatment of the exam topics. Its objective is to prepare you with enough knowledge to pass. As much as practical, this book uses the notation given in the NCEES Handbook.

This book consolidates 180 practical review problems, covering all of the discipline-specific exam topics. All problems include full solutions.

The problems in this book were developed by Robert B. Angus, PE, John E. Hajjar, Abdulrahman Yassine, and Jeffrey Moreland, following the format, style, subject breakdown, and guidelines that I provided.

In developing this book, the NCEES Handbook and the breakdown of problem types published by NCEES were my guide for problem types and scope of coverage. However, as with most standardized tests, there is no guarantee that any specific problem type will be encountered. It is expected that minor variations in problem content will occur from exam to exam.

As with all of PPI's books, the problems in this book are original and have been ethically derived. Although examinee feedback was used to determine its content, this book contains problems that are only *like* those that are on the exam. There are no actual exam problems in this book.

This book was designed to complement my *FE Review Manual*, which you will also need to prepare for the FE exam. The *FE Review Manual* is PPI's most popular study guide for this exam for more than 25 years.

You cannot prepare adequately without your own copy of the NCEES Handbook. This document contains the data and formulas that you will need to solve both the general and the discipline-specific problems. A good way to become familiar with it is to look up the information, formulas, and data that you need while trying to work practice problems.

Exam-prep books are always works in progress. By necessity, a book will change as the exam changes. Even when the exam format doesn't change for a while, new problems and improved explanations can always be added. I encourage you to provide comments via PPI's errata reporting page, **www.ppi2pass.com/errata**. You will find all verified errata there. I appreciate all feedback.

Best of luck to you in your pursuit of licensure.

Michael R. Lindeburg, PE

Engineering Registration in the United States

ENGINEERING REGISTRATION

Engineering registration (also known as *engineering licensing*) in the United States is an examination process by which a state's board of engineering licensing (i.e., registration board) determines and certifies that you have achieved a minimum level of competence. This process protects the public by preventing unqualified individuals from offering engineering services.

Most engineers do not need to be registered. In particular, most engineers who work for companies that design and manufacture products are exempt from the licensing requirement. This is known as the *industrial exemption*. Nevertheless, there are many good reasons for registering. For example, you cannot offer consulting engineering design services in any state unless you are registered in that state. Even within a product-oriented corporation, however, you may find that registered engineers have more opportunities for employment and advancement.

Once you have met the registration requirements, you will be allowed to use the titles Professional Engineer (PE), Registered Engineer (RE), and Consulting Engineer (CE).

Although the registration process is similar in all 50 states, each state has its own registration law. Unless you offer consulting engineering services in more than one state, however, you will not need to register in other states.

The U.S. Registration Procedure

To become a registered engineer in the United States, you will need to pass two eight-hour written examinations. The first is the *Fundamentals of Engineering Examination*, also known as the *Engineer-In-Training Examination* and the *Intern Engineer Exam*. The initials FE, EIT, and IE are also used. This exam covers basic subjects from the mathematics, physics, chemistry, and engineering classes you took during your first four university years. In rare cases, you may be allowed to skip this first exam.

The second eight-hour exam is the *Principles and Practice of Engineering Exam*. The initials PE are also used. This exam is on topics within a specific discipline, and only covers subjects that fall within that area of specialty.

Most states have similar registration procedures. However, the details of registration qualifications, experience requirements, minimum education levels, fees, oral interviews, and exam schedules vary from state to state. For more information, contact your state's registration board (**www.ppi2pass.com/stateboards**).

National Council of Examiners for Engineering and Surveying

The National Council of Examiners for Engineering and Surveying (NCEES) in Clemson, South Carolina, produces, distributes, and scores the national FE and PE exams. The individual states purchase the exams from NCEES and administer them themselves. NCEES does not distribute applications to take the exams, administer the exams or appeals, or notify you of the results. These tasks are all performed by the states.

Reciprocity Among States

With minor exceptions, having a license from one state will not permit you to practice engineering in another state. You must have a professional engineering license from each state in which you work. For most engineers, this is not a problem, but for some, it is. Luckily, it is not too difficult to get a license from every state you work in once you have a license from one state.

All states use the NCEES exams. If you take and pass the FE or PE exam in one state, your certificate will be honored by all of the other states. Although there may be other special requirements imposed by a state, it will not be necessary to retake the FE and PE exams. The issuance of an engineering license based on another state's license is known as *reciprocity* or *comity*.

The simultaneous administration of identical exams in all states has led to the term *uniform examination*. However, each state is still free to choose its own minimum passing score and to add special questions and requirements to the examination process. Therefore, the use of a uniform exam has not, by itself, ensured reciprocity among states.

THE FE EXAM

Applying for the Exam

Each state charges different fees, specifies different requirements, and uses different forms to apply for the exam. Therefore, it will be necessary to request an application from the state in which you want to become registered. Generally, it is sufficient for you to phone for this application. You'll find contact information (websites, telephone numbers, email addresses, etc.) for all U.S. state and territorial boards of registration at **www.ppi2pass.com/stateboards**.

Keep a copy of your exam application, and send the original application by certified mail, requesting a delivery receipt. Keep your proof of mailing and delivery with your copy of the application.

Exam Dates

The national FE and PE exams are administered twice a year (usually in mid-April and late October), on the same weekends in all states. For a current exam schedule, check **www.ppi2pass.com/fefaq**.

FE Exam Format

The NCEES Fundamentals of Engineering examination has the following format and characteristics.

- There are two four-hour sessions separated by a one-hour lunch.

- Examination questions are distributed in a bound examination booklet. A different exam booklet is used for each of the two sessions.

- Formulas and tables of data needed to solve questions in the exams are found in either the NCEES Handbook or in the body of the question statement itself.

- The morning session (also known as the *A.M. session*) has 120 multiple-choice questions, each with four possible answers lettered (A) through (D). Responses must be recorded with a pencil provided by NCEES on special answer sheets. No credit is given for answers recorded in ink.

- Each problem in the morning session is worth one point. The total score possible in the morning is 120 points. Guessing is valid; no points are subtracted for incorrect answers.

- There are questions on the exam from most of the undergraduate engineering degree program subjects. Questions from the same subject are all grouped together, and the subjects are labeled. The percentages of questions for each subject in the morning session are given in the following table.

Morning FE Exam Subjects

subject	percentage of questions (%)
chemistry	9
computers	7
electricity and magnetism	9
engineering economics	8
engineering probability and statistics	7
engineering mechanics (statics and dynamics)	10
ethics and business practices	7
fluid mechanics	7
material properties	7
mathematics	15
strength of materials	7
thermodynamics	7

- There are seven different versions of the afternoon session (also known as the *P.M. session*), six of which correspond to specific engineering disciplines: chemical, civil, electrical, environmental, industrial, and mechanical engineering.

The seventh version of the afternoon exam is a general examination suitable for anyone, but in particular, for engineers whose specialties are not one of the other six disciplines. Though the subjects in the general afternoon exam are similar to the morning subjects, the questions are more complex—hence their double weighting. Questions on the afternoon exam are intended to cover concepts learned in the last two years of a four-year degree program. Unlike morning questions, these questions may deal with more than one basic concept per question.

Each version of the afternoon session consists of 60 questions. All questions are mandatory. Questions in each subject may be grouped into related problem sets containing between two and ten questions each.

The percentages of questions for each subject in the general afternoon session exam are given in the following table.

Afternoon FE Exam Subjects
(Other Disciplines Exam)

subject	percentage of questions (%)
advanced engineering mathematics	10
application of engineering mechanics	13
biology	5
electricity and magnetism	12
engineering economics	10
engineering probability and statistics	9
engineering of materials	11
fluids	15
thermodynamics and heat transfer	15

Each of the discipline-specific afternoon examinations covers a substantially different body of knowledge than the morning exam. The percentages of questions for each subject in the Electrical discipline-specific afternoon session exam are as follows.

Afternoon FE Exam Subjects (Electrical DS Exam)

subject	percentage of questions (%)
circuits	16
power	13
electromagnetics	7
control systems	10
communications	9
signal processing	8
electronics	15
digital systems	12
computer systems	10

Some afternoon questions stand alone, while others are grouped together, with a single problem statement that describes a situation followed by two or more questions about that situation. All questions are multiple-choice. You must choose the best answer from among four, lettered (A) through (D).

- Each question in the afternoon is worth two points, making the total possible score 120 points.
- The scores from the morning and afternoon sessions are added together to determine your total score. No points are subtracted for guessing or incorrect answers. Both sessions are given equal weight. It is not necessary to achieve any minimum score on either the morning or afternoon sessions.
- All grading is done by computer optical sensing.

Use of SI Units on the FE Exam

Metric questions are used in all subjects, except some civil engineering and surveying subjects that typically use only customary U.S. (i.e., English) units. SI units are consistent with ANSI/IEEE standard 268 (the American Standard for Metric Practice). Non-SI metric units might still be used when common or where needed for consistency with tabulated data (e.g., use of bars in pressure measurement).

Grading and Scoring the FE Exam

The FE exam is not graded on the curve, and there is no guarantee that a certain percentage of examinees will pass. Rather, NCEES uses a modification of the Angoff procedure to determine the suggested passing score (the cutoff point or cut score).

With this method, a group of engineering professors and other experts estimate the fraction of minimally qualified engineers who will be able to answer each question correctly. The summation of the estimated fractions for all test questions becomes the passing score. Because the law in most states requires engineers to achieve a score of 70% to become licensed, you may be reported as having achieved a score of 70% if your raw score is greater than the passing score established by NCEES, regardless of the raw percentage. The actual score may be slightly more or slightly less than 110 as determined from the performance of all examinees on the equating subtest.

About 20% of the FE exam questions are repeated from previous exams—this is the *equating subtest*. Since the scores of previous examinees on the equating subtest are known, comparisons can be made between the two exams and examinee populations. These comparisons are used to adjust the passing score.

The individual states are free to adopt their own passing score, but all adopt NCEES's suggested passing score because the states believe this cutoff score can be defended if challenged.

You will receive the results within 12 weeks of taking the exam. If you pass, you will receive a letter stating that you have passed. If you fail, you will be notified that you failed and be provided with a diagnostic report.

Permitted Reference Material

Since October 1993, the FE exam has been what NCEES calls a "limited-reference" exam. This means that no books or references other than those supplied by NCEES may be used. Therefore, the FE exam is really an "NCEES-publication only" exam. NCEES provides its own Supplied-Reference Handbook for use during the examination. No books from other publishers may be used.

CALCULATORS

In most states, battery- and solar-powered, silent calculators can be used during the exam, although printers cannot be used. (Solar-powered calculators are preferable because they do not have batteries that run down.) In most states, programmable, preprogrammed, and business/finance calculators are allowed. Contact your state board to determine if nomographs and slide rules are permitted. To prevent unauthorized transcription and redistribution of the exam questions, calculators with communication or text-editing capabilities are banned from all NCEES exam sites. You cannot share calculators with other examinees. For a list of allowed calculators check **www.ppi2pass.com/calculators**.

It is essential that a calculator used for engineering examinations have the following functions.

- trigonometric functions
- inverse trigonometric functions
- hyperbolic functions
- pi
- square root and x^2
- common and natural logarithms
- y^x and e^x

For maximum speed, your calculator should also have or be programmed for the following functions.

- extracting roots of quadratic and higher-order equations
- converting between polar (phasor) and rectangular vectors
- finding standard deviations and variances
- calculating determinants of 3×3 matrices
- linear regression
- economic analysis and other financial functions

STRATEGIES FOR PASSING THE FE EXAM

The most successful strategy for passing the FE exam is to prepare in all of the exam subjects. Do not limit the number of subjects you study in hopes of finding enough questions in your strongest areas of knowledge to pass.

Fast recall and stamina are essential to doing well. You must be able to quickly recall solution procedures, formulas, and important data. You will not have time during the exam to derive solutions methods—you must know them instinctively. This ability must be maintained for eight hours. Be sure to gain familiarity with the NCEES Handbook by using it as your only reference for some of the problems you work during study sessions.

In order to get exposure to all exam subjects, it is imperative that you develop and adhere to a review schedule. If you are not taking a classroom review course (where the order of your preparation is determined by the lectures), prepare your own review schedule.

There are also physical demands on your body during the exam. It is very difficult to remain alert and attentive for eight hours or more. Unfortunately, the more time you study, the less time you have to maintain your physical condition. Thus, most examinees arrive at the exam site in peak mental condition but in deteriorated physical condition. While preparing for the FE exam is not the only good reason for embarking on a physical conditioning program, it can serve as a good incentive to get in shape.

It will be helpful to make a few simple decisions prior to starting your review. You should be aware of the different options available to you. For example, you should decide early on to

- use SI units in your preparation
- perform electrical calculations with effective (rms) or maximum values
- take calculations out to a maximum of four significant digits
- prepare in all exam subjects, not just your specialty areas

At the beginning of your review program, you should locate a spare calculator. It is not necessary to buy a spare if you can arrange to borrow one from a friend or the office. However, if possible, your primary and spare calculators should be identical. If your spare calculator is not identical to the primary calculator, spend some time familiarizing yourself with its functions.

A Few Days Before the Exam

There are a few things you should do a week or so before the exam date. For example, visit the exam site in order to find the building, parking areas, examination room, and rest rooms. You should also make arrangements now for child care and transportation. Since the exam does not always start or end at the designated times, make sure that your child care and transportation arrangements can tolerate a late completion.

Next in importance to your scholastic preparation is the preparation of your two examination kits. The first kit consists of a bag or box containing items to bring with you into the examination room.

[ ] letter admitting you to the exam
[ ] photographic identification
[ ] main calculator
[ ] spare calculator
[ ] extra calculator batteries
[ ] unobtrusive snacks
[ ] travel pack of tissues
[ ] headache remedy
[ ] $2.00 in change
[ ] light, comfortable sweater
[ ] loose shoes or slippers
[ ] handkerchief
[ ] cushion for your chair
[ ] small hand towel
[ ] earplugs
[ ] wristwatch with alarm
[ ] wire coat hanger
[ ] extra set of car keys

The second kit consists of the following items and should be left in a separate bag or box in your car in case you need them.

- [ ] copy of your application
- [ ] proof of delivery
- [ ] this book
- [ ] other references
- [ ] regular dictionary
- [ ] scientific dictionary
- [ ] course notes in three-ring binders
- [ ] instruction booklets for all your calculators
- [ ] light lunch
- [ ] beverages in thermos and cans
- [ ] sunglasses
- [ ] extra pair of prescription glasses
- [ ] raincoat, boots, gloves, hat, and umbrella
- [ ] street map of the exam site
- [ ] note to the parking patrol for your windshield explaining where you are, what you are doing, and why your time may have expired
- [ ] battery-powered desk lamp

The Day Before the Exam

Take the day before the exam off from work to relax. Do not cram the last night. A good prior night's sleep is the best way to start the exam. If you live far from the exam site, consider getting a hotel room in which to spend the night.

Make sure your exam kits are packed and ready to go.

The Day of the Exam

You should arrive at least 30 minutes before the exam starts. This will allow time for finding a convenient parking place, bringing your materials to the exam room, and making room and seating changes. Be prepared, though, to find that the examination room is not open or ready at the designated time.

Once the examination has started, consider the following suggestions.

- Set your wristwatch alarm for five minutes before the end of each four-hour session, and use that remaining time to guess at all of the remaining unsolved problems. Do not work up until the very end. You will be successful with about 25% of your guesses, and these points will more than make up for the few points you might earn by working during the last five minutes.

- Do not spend more than two minutes per morning question. (The average time available per problem is two minutes.) If you have not finished a question in that time, make a note of it and move on.

- Do not ask your proctors technical questions. Even if they are knowledgeable in engineering, they will not be permitted to answer your questions.

- Make a quick mental note about any problems for which you cannot find a correct response or for which you believe there are two correct answers. Errors in the exam are rare, but they do occur. Being able to point out an error later might give you the margin you need to pass. Since such problems are almost always discovered during the scoring process and discounted from the exam, it is not necessary to tell your proctor, but be sure to mark the one best answer before moving on.

- Make sure all of your responses on the answer sheet are dark and completely fill the bubbles.

Common Questions About the DS Exam

Q: Do I have to take the DS exam?

A: Most people do not have to take the DS exam and may elect the general exam option. The state boards do not care which afternoon option you choose, nor do employers. In some cases, an examinee still in an undergraduate degree program may be required by his or her university to take a specific DS exam.

Q: Do all mechanical, civil, electrical, chemical, industrial, and environmental engineers take the DS exam?

A: Originally, the concept was that examinees from the "big five" disciplines would take the DS exam, and the general exam would be for everyone else. This remains just a concept, however. A majority of engineers in all of the disciplines apparently take the general exam.

Q: When do I elect to take the DS exam?

A: You will make your decision when registering for the FE exam.

Q: When do I choose which DS exam I want to take?

A: You must specify your desired DS exam when registering for the FE exam. Each topic is a separate booklet and you will receive only the exam indicated on your application.

Q: After I take the DS exam, does anyone know that I took it?

A: After you take the FE exam, only NCEES and your state board will know whether you took the DS or general exam. Such information may or may not be retained by your state board.

Q: Will my DS FE certificate be recognized by other states?

A: Yes. All states recognize passing the FE exam and do not distinguish between the DS and general afternoon portions of the FE exam.

Q: Is the DS FE certificate "better" than the general FE certificate?

A: There is no difference. No one will know which option you chose. It's not stated on the certificate you receive from your state.

Q: What is the format of the DS exam?

A: The DS exam is 4 hours long. There are 60 problems, each worth 2 points. The average time per problem is 4 minutes. Each problem is multiple choice with 4 answer choices. Most problems require the application of more than one concept (i.e., formula).

Q: Is there anything special about the way the DS exam is administered?

A: In all ways, the DS and general afternoon exam are equivalent. There is no penalty for guessing. No credit is given for scratch pad work, methods, etc.

Q: Are the answer choices close or tricky?

A: Answer choices are not particularly close together in value, so the number of significant digits is not going to be an issue. Wrong answers, referred to as "distractors" by NCEES, are credible. However, the exam is not "tricky"; it does not try to mislead you.

Q: Are any problems in the afternoon session related to each other?

A: Several questions may refer to the same situation or figure. However, NCEES has tried to make all of the questions independent. If you make a mistake on one question, it shouldn't carry over to another.

Q: Is there any minimum passing score for the DS exam?

A: No. It is the total score from your morning and afternoon sessions that determines your passing, not the individual session scores. You do not have to "pass" each session individually.

Q: Is the general portion easier, harder, or the same as the DS exams?

A: Theoretically, all of the afternoon options are the same. At least, that is the intent of offering the specific options—to reduce the variability. Individual passing rates, however, may still vary 5% to 10% from exam to exam. (PPI lists the most recent passing statistics for the various DS options on its website at **www.ppi2pass.com/fepassrates**.)

Q: Do the DS exams cover material at the undergraduate or graduate level?

A: Like the general exam, test topics come entirely from the typical undergraduate degree program. However, the emphasis is primarily on material from the third and fourth year of your program. This may put examinees who take the exam in their junior year at a disadvantage.

Q: Do you need practical work experience to take the DS exam?

A: No.

Q: Does the DS exam also draw on subjects that are in the general exam?

A: Yes. The dividing line between general and DS topics is often indistinct.

Q: Is the DS exam in customary U.S. or SI units?

A: The DS exam is nearly entirely in SI units. A few exceptions exist for some engineering subjects (surveying, hydrology, code-based design, etc.) where current common practice uses only customary U.S. units.

Q: Does the NCEES Handbook cover everything that is on the DS exam?

A: No. You may be tested on subjects that are not in the NCEES Handbook. However, NCEES has apparently adopted an unofficial policy of providing any necessary information, data, and formulas in the stem of the question. You will not be required to memorize any formulas.

Q: Is everything in the DS portion of the NCEES Handbook going to be on the exam?

A: Apparently, there is a fair amount of reference material that isn't needed for every exam. There is no way, however, to know in advance what material is needed.

Q: How long does it take to prepare for the DS exam?

A: Preparing for the DS exam is similar to preparing for a mini PE exam. Engineers typically take two to four months to complete a thorough review for the PE exam. However, examinees who are still in their degree program at a university probably aren't going to spend more than two weeks thinking about, worrying about, or preparing for the DS exam. They rely on their recent familiarity with the subject matter.

Q: If I take the DS exam and fail, do I have to take the DS exam the next time?

A: No. The examination process has no memory.

Q: Where can I get even more information about the DS exam?

A: Visit the Exam FAQs and the Engineering Exam Forum at PPI's website (**www.ppi2pass.com/fe**).

How to Use This Book

HOW EXAMINEES CAN USE THIS BOOK

This book is divided into three parts: The first part consists of 60 representative practice problems covering all of the topics in the afternoon DS exam. Sixty problems corresponds to the number of problems in the afternoon DS exam. You may time yourself by allowing approximately 4 minutes per problem when attempting to solve these problems, but that was not my intent when designing this book. Since the solution follows directly after each problem in this section, I intended for you to read through the problems, attempt to solve them on your own, become familiar with the support material in the official NCEES Handbook, and accumulate the reference materials you think you will need for additional study.

The second and third parts of this book consists of two complete sample examinations that you can use as sources of additional practice problems or as timed diagnostic tools. They also contain 60 problems, and the number of problems in each subject corresponds to the breakdown of subjects published by NCEES. Since the solutions to these parts of the book are consolidated at the end, it was my intent that you would solve these problems in a realistic mock-exam mode.

You should use the NCEES Handbook as your only reference during the mock exams.

The morning general exam and the afternoon DS exam essentially cover two different bodies of knowledge. It takes a lot of discipline to prepare for two standardized exams simultaneously. Because of that (and because of my good understanding of human nature), I suspect that you will be tempted to start preparing for your chosen DS exam only after you have become comfortable with the general subjects. That's actually quite logical, because if you run out of time, you will still have the general afternoon exam as a viable option.

If, however, you are limited in time to only two or three months of study, it will be quite difficult to do a thorough DS review if you wait until after you have finished your general review. With a limited amount of time, you really need to prepare for both exams in parallel.

HOW INSTRUCTORS CAN USE THIS BOOK

The availability of the discipline-specific FE exam has greatly complicated the lives of review course instructors and coordinators. The general consensus is that it is essentially impossible to do justice to all of the general FE exam topics and then present a credible review for each of the DS topics. Increases in course cost, expenses, course length, and instructor pools (among many other issues) all conspire to create quite a difficult situation.

One-day reviews for each DS subject are subject-overload from a reviewing examinee's standpoint. Efforts to shuffle FE students over the parallel PE review courses meet with scheduling conflicts. Another idea, that of lengthening lectures and providing more in-depth coverage of existing topics (e.g., covering transistors during the electricity lecture), is perceived as a misuse of time by a majority of the review course attendees. Is it any wonder that virtually every FE review course in the country has elected to only present reviews for the general afternoon exam?

But, while more than half of the examinees elect to take the other disciplines afternoon exam, some may actually be required to take a DS exam. This is particularly the case in some university environments where the FE exam has become useful as an "outcome assessment tool." Thus, some method of review is still needed.

Since most examinees begin reviewing approximately two to three months before the exam (which corresponds to when most review courses begin), it is impractical to wait until the end of the general review to start the DS review. The DS review must proceed in parallel with the general review.

In the absence of parallel DS lectures (something that isn't yet occurring in too many review courses), you may want to structure your review course to provide lectures only on the general subjects. Your DS review could be assigned as "independent study," using chapters and problems from this book. Thus, your DS review would consist of distributing this book with a schedule of assignments. Your instructional staff could still provide assistance on specific DS problems, and completed DS assignments could still be recorded.

The final chapter on incorporating DS subjects into review courses has yet to be written. Like the landscape architect who waits until a well-worn path appears through the plants before placing stepping stones, we need to see how review courses do it before we can give any advice.

Nomenclature

NOMENCLATURE

a	number system digit	–
a	unit vector in x direction	–
A	amplitude	–
A	area	m^2
A	Fourier even coefficient	–
A	system matrix	–
b	binary bit (1 or 0)	–
B	bandwidth	Hz
B	Fourier odd coefficient	–
B	input distribution matrix	–
B	magnetic flux density	T
C	capacitance	F
C	Fourier complex coefficient	–
C	output matrix	–
$C(s)$	control system controlled output	–
D	decimal equivalent	–
D	electric displacement	C/m^2
D	feed-through matrix	–
$D(s)$	denominator polynomial	–
E	electric field strength	V/m
f	frequency	Hz
F	force	N
$f(t)$	function of time	–
G	gain	–
g_m	transconductance	S
$G(s)$	partitioned plant gain	–
$G_C(s)$	controller or compensator gain	–
GF	gage factor	–
GM	gain margin	–
$G_R(s)$	input processor gain	–
G_{ss}	steady-state gain	–
H	magnetic field strength	A/m
$H(s)$	feedback gain	–
$H(s)$	transfer function	–
i	instantaneous current	A
i	number (integer)	–
I	DC or effective current	A
I	identity matrix (1 on diagonal, 0 elsewhere)	–
J	current density	A/m^2
k	Boltzman's constant	1.38×10^{-23} J/K
k	number (integer)	–
K	conductivity factor	A/V^2
K	constant	–
K	control system steady-state gain	–
K	number (integer)	–
K_n	noise gain	–
k_f	frequency angle modulation amplitude	–
k_p	phase angle modulation amplitude	–
K_B	constant gain for control systems inputs	–
K_c	constant controller amplitude	–
l	length	m
L	inductance	H
L	length	m
$L(s)$	control system load disturbance	–
m	index of AM modulation	–
m	modulation amplitude	–
m	number (integer)	–
M	mantissa	–
M	N-digit, r-base number	–
M	number (integer)	–
$m(t)$	modulation baseband signal	various
M_p	magnitude of second-order system peak	various
n	electron concentration	$1/m^3$
n	number (integer)	–
n	rotational speed	rev/s
N	concentration	$1/m^3$
N	number (integer)	–
n_i	intrinsic carrier concentration	$1/m^3$
n_n	electron concentration in n-type material	$1/m^3$
$N(s)$	numerator polynomial	–
p	hole concentration	$1/m^3$
p	number of poles (motor)	–
P	power (real)	W
pf	power factor	–
p_p	hole concentration in p-type material	$1/m^3$
PM	phase margin	degrees
POS	product of sums	–
PRF	pulse repetition frequency	–
q	charge of an electron	1.602×10^{-19} C

Q	charge	C
Q	power (reactive)	VAR
r	base of a number system	–
r	distance	m
r	electrical resistance	Ω
R	electrical resistance	Ω
R	radix complement	–
r_b	BJT base resistance	Ω
r_d	FET drain-source resistance	Ω
r_o	BJT equivalent collector-base resistance	Ω
$r(t)$	correlation of nonperiodic functions $x(t)$ and $y(t)$	–
$R_\Box$	sheet electrical resistance	Ω
$R(s)$	control system set point	–
s	Laplace frequency domain variable	1/s
s	slip	–
S	apparent (complex) power	VA
S	sign bit	–
S	surface area	m^2
$\mathbf{S}$	surface area vector	–
SWR	standing wave ratio	–
t	time	s
T	period of oscillation	s
T	temperature	K or °C
T	thickness	m
T	torque	N·m
T	transient gain	–
T	type of control system exponent	–
T_D	gain for derivative	–
T_I	gain for integral	–
T_S	settling time	s
u	magnetic energy	J
$\mathbf{u}(t)$	R by 1 input vector (R inputs)	–
$\mathbf{U}(s)$	Laplace transform of input vector	–
v	velocity	m/s
v	voltage	V
v	volume	m^3
V	DC or effective voltage	V
V	volume	m^3
$v(t)$	convolution of $x(t)$ and $y(t)$	–
V_a	early voltage	V
V_{bi}	junction contact potential	V
w	uncertainty (of a measurement)	various
W	width	m
$\mathbf{x}(t)$	N by 1 state vector (N state variables)	–
$\mathbf{X}(s)$	Laplace transform of state vector	–
$\mathbf{y}(t)$	M by 1 output vector (M outputs)	–
$\mathbf{Y}(s)$	output vector Laplace transform	–
Z	impedance	Ω

SYMBOLS

α	collector-emitter current ratio	–
β	collector-base current ratio	–
β	propagation constant	–
Γ	reflection coefficient	–
δ	Dirac delta function	–
δ	logarithmic decrement	–
ε	normal strain on a strain gage	–
ε	permittivity	F/m
ε_p	position error coefficient	–
ε_r	permittivity ratio (dielectric constant)	–
ε_o	permittivity of free space	8.854×10^{-12} F/m
ζ	damping ratio	–
η	emission coefficient	–
θ_P	power factor angle	degrees or radians
λ	wavelength	m
μ	mobility	m^2/V·s
μ	permeability	H/m
μ_0	permeability of free space	$4\pi \times 10^{-7}$ H/m
μ_n	mobility of electrons	m^2/V·s
μ_p	mobility of holes	m^2/V·s
ρ	charge density	C/m^3
ρ	reflection coefficient	–
ρ	resistivity	Ω·m
σ	conductivity	S/m
τ	dummy variable	–
τ	period of oscillation	s
τ	time constant	s
ϕ	magnetic flux	Wb
$\phi(t)$	phase angle modulation	radians
$\phi(t)$	state transition matrix	–
$\Phi(s)$	resolution matrix	–
ω	angular frequency	rad/s

SUBSCRIPTS

$\Box$	sheet
a	acceptor
a	armature
a	donor atoms
a	phase a
ab	phase a to phase b
an	phase a to neutral
b	phase b
B	base

B	breakdown	p	damped resonant
bc	phase b to phase c	p	hole
BE	base-emitter	p	pinch-off
BEsat	base-emitter maximum (saturation)	p	primary
bn	phase b to neutral	pp	peak-to-peak
c	carrier	p	phase
c	center	r	relative
c	phase c	rms	root-mean-square
C	capacitor	s	sampling
C	collector	s	secondary
ca	phase c to phase a	s	settling
CE	collector-emitter	s	shunt
CEsat	collector-emitter maximum (saturation)	s	sideband
CL	closed loop	s	synchronous
cn	phase c to neutral	s	source
d	damped natural	S	saturation
d	donor	S	source
d	drain	S	surface
d	impurities	t	threshold
D	diode	t	total
D	drain	t	transmitted
ds	drain-source	T	thermal
DS	drain-source	Th	Thevenin equivalent
DSS	drain-source saturation	V	volume
e	electron		
E	emitter		
encl	enclosed		
f	feedback		
f	field		
f	frequency		
fs	full scale		
g	generator		
G	ground		
GD	gate drain		
GS	gate-source		
h	high		
i	instantaneous		
i	intrinsic		
K	conductivity factor		
l	low		
L	line		
L	load		
LN	line-to-neutral		
m	internal		
m	mechanical		
m	meter		
m	modulation		
m	number (integer)		
n	natural		
n	noise		
n	number (integer)		
o	characteristic		
o	initial		
o	output		
oc	open circuit		

Practice Problems

CIRCUITS

Problem 1

The circuit shown is to be used as a filter. The cutoff frequency, f_c, of each R and C pair of components is computed using the following formula.

$$f_c = \frac{1}{2\pi RC}$$

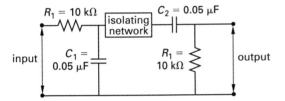

This circuit is a

(A) low-pass filter
(B) high-pass filter
(C) band-reject filter
(D) band-pass filter

Solution

R_1 and C_1 act as a low-pass filter. The cutoff frequency is

$$\begin{aligned}f_c &= \frac{1}{2\pi R_1 C_1} \\ &= \frac{1}{(2\pi)(10 \times 10^3 \; \Omega)(0.05 \times 10^{-6} \; \text{F})} \\ &= 318.3 \; \text{Hz}\end{aligned}$$

The isolating network prevents R_2 and C_2 from loading R_1 and C_1, while R_2 and C_2 act as a high-pass filter whose cutoff frequency is also 318.3 Hz.

The combination acts as a band-pass filter because all frequencies significantly above 318.3 Hz and below 318.3 Hz are attenuated.

The answer is D.

Problem 2

For a voltage gain of 40,000, what is the equivalent decibel gain?

(A) 40
(B) 46
(C) 92
(D) 100

Solution

The equation relating gain in dB and the voltage gain as a ratio is

$$\begin{aligned}\text{gain} &= 20 \; \log_{10} K \\ &= 20 \; \log_{10}(40{,}000) \\ &= 92.04 \; \text{dB} \quad (92 \; \text{dB})\end{aligned}$$

The answer is C.

Problem 3

What is the time constant for the circuit shown?

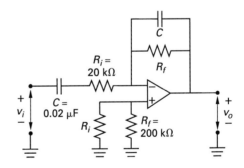

(A) 2.0 ms
(B) 4.0 ms
(C) 25 ms
(D) 100 ms

Solution

The time constant of the circuit is the product of R_f and C.

$$\begin{aligned}\tau = R_f C &= (200 \times 10^3 \; \Omega)(0.02 \times 10^{-6} \; \text{F}) \\ &= 4.0 \times 10^{-3} \; \text{s} \quad (4.0 \; \text{ms})\end{aligned}$$

The answer is B.

Problem 4
The circuit shown is an astable (free-running) multivibrator.

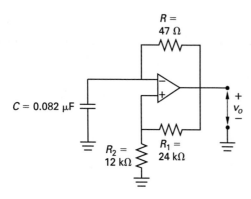

What is the feedback voltage for this circuit if v_o is 11.5 V?

(A) 2.3 V
(B) 2.9 V
(C) 3.8 V
(D) 5.8 V

Solution
The feedback voltage is

$$v_f = \left(\frac{R_2}{R_1 + R_2}\right) v_o$$

$$= \left(\frac{12 \times 10^3 \ \Omega}{24 \times 10^3 \ \Omega + 12 \times 10^3 \ \Omega}\right)(11.5 \text{ V})$$

$$= 3.833 \text{ V} \quad (3.8 \text{ V})$$

The answer is C.

Problem 5
The following circuit is used to measure BJT transistor parameters.

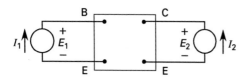

If $E_2 = 0$ V, $E_1 = 2$ V, $I_1 = 0.5$ mA, and $I_2 = 100$ mA, what is the value of the collector-base current ratio, β?

(A) 100
(B) 200
(C) 400
(D) 500

Solution
Because $E_2 = 0$,

$$\beta \approx \frac{I_C}{I_B} = \frac{I_2}{I_1}$$

$$= \frac{100 \text{ mA}}{0.5 \text{ mA}}$$

$$= 200$$

The answer is B.

Problem 6
At $t = 0^-$ s, the circuit given in the following illustration is stable. At $t = 0$ s, a step of 40 V is applied as indicated. What is the voltage across the capacitor at $t = 0^-$ s?

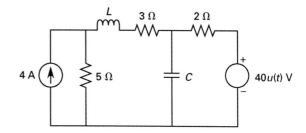

(A) 0 V
(B) 4 V
(C) 20 V
(D) 40 V

Solution
Prior to $t = 0$ s, the step-voltage source is equivalent to a short circuit, the inductor is equivalent to a short circuit, and the capacitor is equivalent to an open circuit. Thus the equivalent circuit at $t = 0^-$ s is shown as follows.

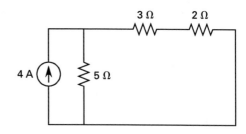

The 4 A current distributes equally between the 5 Ω in parallel and the two series resistors. The capacitor voltage is also the 2 Ω resistor voltage. The capacitor voltage, by Ohm's law, is

$$V_C = I_C R = (2 \text{ A})(2 \ \Omega)$$

$$= 4 \text{ V}$$

The answer is B.

Problem 7

For the circuit given, what is the final voltage across the capacitor after the switch is closed?

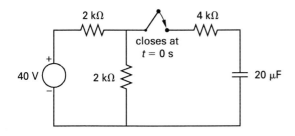

(A) 10 V
(B) 20 V
(C) 30 V
(D) 40 V

Solution

At the end of the transient period (after the switch is closed), the capacitor acts like an open circuit. There is no current through the open-circuit capacitor. Therefore, the voltage across the capacitor is the same as the voltage across the vertical 2 kΩ resistor. Apply the voltage-divider formula.

$$V_C = (40 \text{ V})\left(\frac{2 \text{ k}\Omega}{2 \text{ k}\Omega + 2 \text{ k}\Omega}\right)$$
$$= 20 \text{ V}$$

The answer is B.

Problem 8

What is the time constant for the circuit shown when the switch is closed?

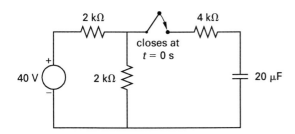

(A) 80 ms
(B) 100 ms
(C) 160 ms
(D) 320 ms

Solution

Because the switch acts as a short circuit when the voltage source is 0 V, the pair of 2 kΩ resistors will be connected in parallel with each other and in series with the 4 kΩ resistor. The equivalent resistance across the capacitor is

$$R_{equiv} = 4 \text{ k}\Omega + \frac{(2 \text{ k}\Omega)(2 \text{ k}\Omega)}{2 \text{ k}\Omega + 2 \text{ k}\Omega}$$
$$= 4 \text{ k}\Omega + 1 \text{ k}\Omega$$
$$= 5 \text{ k}\Omega$$

The time constant is the product of this R_{equiv} and the capacitance value.

$$\tau = R_{equiv} C = (5 \times 10^3 \Omega)(20 \times 10^{-6} \text{ F})$$
$$= 0.1 \text{ s} \quad (100 \text{ ms})$$

The answer is B.

Problem 9

The switch in the following circuit closes at $t = 0$. What is the approximate current through the inductor 60 μs after the switch is closed?

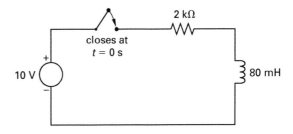

(A) 2.4 mA
(B) 3.2 mA
(C) 3.9 mA
(D) 5.0 mA

Solution

The equation for transient inductor current is

$$i_L = \left(\frac{V}{R}\right)\left(1 - e^{-\frac{Rt}{L}}\right)$$
$$= \left(\frac{10 \text{ V}}{2 \times 10^3 \Omega}\right)\left(1 - e^{-\frac{(2 \times 10^3 \Omega)(60 \times 10^{-6} \text{ s})}{80 \times 10^{-3} \text{ H}}}\right)$$
$$= 3.88 \times 10^{-3} \text{ A} \quad (3.9 \text{ mA})$$

The answer is C.

Problem 10

What are the poles of the following voltage function?

$$V(s) = \frac{s^2 + 7s + 12}{s^3 + 7s^2 + 10s}$$

(A) −3, −4
(B) 0, −3, −4
(C) −2, −5
(D) 0, −2, −5

Solution

$$V(s) = \frac{(s+3)(s+4)}{s(s+2)(s+5)}$$

The poles occur where the individual denominator factors become zero.

$$s = 0$$
$$s + 2 = 0$$
$$s = -2$$
$$s + 5 = 0$$
$$s = -5$$

The answer is D.

Problem 11

The switch for the following circuit is closed at $t = 0$ s. There is an initial charge of 10 V on the capacitor. What is the current just after the switch is closed?

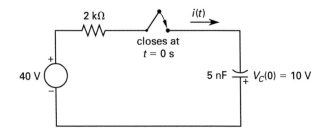

(A) 15 mA
(B) 20 mA
(C) 25 mA
(D) 30 mA

Solution

At $t = 0$ s (when the switch is closed), the capacitor equivalent circuit is a short circuit in series with the 10 V charge. Thus, the circuit current at $t = 0$ s can be calculated realizing that the two voltage sources are in series and combine linearly.

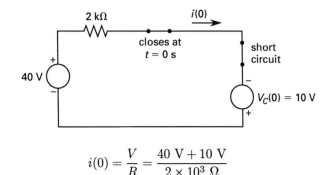

$$i(0) = \frac{V}{R} = \frac{40 \text{ V} + 10 \text{ V}}{2 \times 10^3 \text{ }\Omega}$$
$$= 25 \times 10^{-3} \text{ A} \quad (25 \text{ mA})$$

The answer is C.

POWER

Problem 12

A 500 hp, three-phase, delta-connected induction motor draws 100 A on each of its terminals under full load. The power source is a wye-connected transformer whose line-to-neutral voltage is 2400 V. Assuming the motor efficiency is 100%, the motor's power factor is most nearly

(A) 0.52
(B) 0.64
(C) 0.80
(D) 0.90

Solution

A three-phase motor is a balanced load. Since it is connected in a delta configuration, the line and phase voltages are equivalent. The phase currents are related to the line currents by

$$I_L = \sqrt{3}\, I_p$$

The total power is three times the per-phase power

$$P_t = 3P_p = 3I_p V_L \cos\theta$$
$$= (3)\left(\frac{I_L}{\sqrt{3}}\right) V_L \cos\theta$$
$$= \sqrt{3} I_L V_L \cos\theta$$

The line voltage from the transformer is found by multiplying its line-to-neutral voltage by $\sqrt{3}$.

$$V_L = \sqrt{3} V_{\text{LN}} = (\sqrt{3})(2400 \text{ V})$$
$$= 4157 \text{ V}$$

The total power is

$$P_t = (\sqrt{3})(100 \text{ A})(4157 \text{ V}) \cos\theta$$
$$= 720\,013 \text{ VA} \cos\theta$$

The power factor, $\cos\theta$, is

$$\cos\theta = \frac{P_t}{S} = \frac{(500 \text{ hp})\left(746\, \frac{\text{W}}{\text{hp}}\right)}{720\,013 \text{ VA}}$$
$$= 0.518 \quad (0.52)$$

The answer is A.

Problem 13
The circuit shown has a power factor of 0.8. What size capacitor must be added to this circuit to correct the power factor to 0.9?

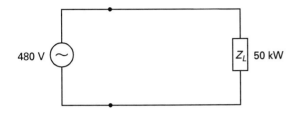

(A) 10 kVAR in series
(B) 10 kVAR in parallel
(C) 13 kVAR in series
(D) 13 kVAR in parallel

Solution
The power triangle for the system is shown as

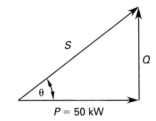

$$\text{pf} = \cos\theta = 0.8$$
$$S = \frac{P}{\text{pf}} = \frac{50 \text{ kW}}{0.8}$$
$$= 62.5 \text{ kVA}$$
$$Q = \sqrt{S^2 - P^2}$$
$$= \sqrt{(62.5 \text{ kVA})^2 - (50 \text{ kW})^2}$$
$$= 37.5 \text{ kVAR}$$

The power triangle for the corrected system is shown as

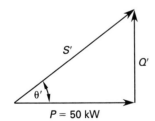

$$\text{pf} = \cos\theta$$
$$= 0.9$$
$$S' = \frac{P}{\text{pf}} = \frac{50 \text{ kW}}{0.9}$$
$$= 55.56 \text{ kVA}$$

$$Q' = \sqrt{S'^2 - P^2}$$
$$= \sqrt{(55.56 \text{ kVA})^2 - (50 \text{ kW})^2}$$
$$= 24.2 \text{ kVAR}$$

Connecting the capacitor in series with the load allows its reactance to be added directly.

$$Q' = Q + Q_{\text{capacitor}}$$
$$Q_{\text{capacitor}} = Q' - Q$$
$$= 24.2 \text{ kVAR} - 37.5 \text{ kVAR}$$
$$= -13.3 \text{ kVAR} \quad [\text{leading}]$$

The answer is C.

Problem 14
In the circuit shown, load Z_L has a power factor of 0.8. What is the overall system power factor?

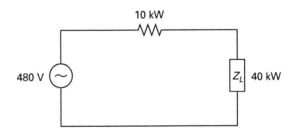

(A) 0.50
(B) 0.80
(C) 0.86
(D) 0.92

Solution
For the load Z_L, the power triangle is shown by

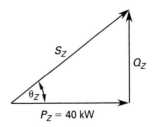

$$\text{pf} = \cos\theta_Z$$
$$\theta_Z = \cos^{-1} 0.8 = 36.87°$$
$$Q_Z = P_Z \tan\theta_z = 40 \text{ kW} \tan 36.87°$$
$$= 30 \text{ kVAR}$$

For the overall system, the power triangle is given by

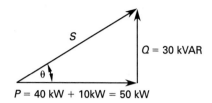

$$\theta = \tan^{-1}\frac{Q}{P}$$
$$= \tan^{-1}\left(\frac{30 \text{ kVAR}}{50 \text{ kW}}\right)$$
$$= 31°$$
$$\text{pf} = \cos\theta = \cos 31°$$
$$= 0.86$$

The answer is C.

Problem 15
Two loads are connected in parallel to a 120 V source. The first load dissipates 5 kW and absorbs 5 kVAR (lagging). The second load dissipates 10 kW and absorbs 10 kVAR (leading). The impedance seen by the 120 V source is most nearly

(A) $0.64\,\Omega\angle-18°$
(B) $0.64\,\Omega\angle 18°$
(C) $0.91\,\Omega\angle-18°$
(D) $0.96\,\Omega\angle 45°$

Solution
The first load has an impedance given by

$$|Z_1| = \frac{V_L^2}{S_1} = \frac{V_L^2}{\sqrt{P_1^2+Q_1^2}}$$
$$= \frac{(120 \text{ V})^2}{\sqrt{(5\times 10^3 \text{ W})^2 + (5\times 10^3 \text{ VAR})^2}}$$
$$= 2.036\,\Omega$$

$$\angle Z_1 = \angle S_1 = \tan^{-1}\frac{Q_1}{P_1}$$
$$= \tan^{-1}\left(\frac{5 \text{ kVAR}}{5 \text{ kW}}\right)$$
$$= 45° \quad \text{[lagging]}$$

The second load has an impedance given by

$$|Z_2| = \frac{V_L^2}{S_2} = \frac{V_L^2}{\sqrt{P_2^2+Q_2^2}}$$
$$= \frac{(120 \text{ V})^2}{\sqrt{(10\times 10^3 \text{ W})^2 + (10\times 10^3 \text{ VAR})^2}}$$
$$= 1.018\,\Omega$$

$$\angle Z_2 = \angle S_2 = \tan^{-1}\frac{Q_2}{P_2}$$
$$= \tan^{-1}\left(\frac{-10 \text{ kVAR}}{10 \text{ kW}}\right)$$
$$= -45° \quad \text{[leading]}$$

The total impedance is given by

$$Z = Z_1 \| Z_2$$
$$= \frac{(2.036\,\Omega\angle 45°)(1.018\,\Omega\angle -45°)}{2.036\,\Omega\angle 45° + 1.018\,\Omega\angle -45°}$$
$$= 0.9105\,\Omega\angle -18.4° \quad (0.91\,\Omega\angle -18°)$$

The answer is C.

Problem 16
For a balanced 240 V, AC three-phase system, load A = 12 kW at a 0.7 lagging power factor, and load B = 18 kW at a 0.8 leading power factor. What are the total real and reactive powers delivered to the loads by the source?

(A) 30 kW, 25.8 kVAR leading
(B) 6 kW, 25.8 kVAR lagging
(C) 30 kW, 1.26 kVAR leading
(D) 30 kW, 1.26 kVAR lagging

Solution
Load A has a positive (lagging) power-factor angle of $\arccos 0.7 = 45.57°$. The real power is 12 kW; the reactive power is

$$(12 \text{ kW})\tan 45.57° = 12.24 \text{ kVAR}$$

Load B has a negative (leading) power-factor angle of $-\arccos 0.8 = -36.87°$. The real power is 18 kW; the reactive power is

$$(18 \text{ kW})\tan -36.87° = -13.50 \text{ kVAR}$$

The total real power is

$$12 \text{ kW} + 18 \text{ kW} = 30 \text{ kW}$$

The total reactive power is

$$12.24 \text{ kVAR} - 13.50 \text{ kVAR} = -1.26 \text{ kVAR}$$
$$\text{[leading]}$$

The answer is C.

Problem 17
The flux that links the primary and secondary coils of a transformer varies according to

$$\phi_{12}(t) = 5 \text{ mWb} \sin 377t$$

The secondary coil contains 500 turns. What is most nearly the secondary voltage, V_2?

(A) 2.5 V sin 377t
(B) 2.5 V cos 377t
(C) 940 V sin 377t
(D) 940 V cos 377t

Solution

The secondary voltage is

$$V_2 = N_2 \frac{d\phi_{12}}{dt}$$
$$= (500 \text{ turns}) \left(\frac{d((5 \times 10^{-3} \text{ Wb}) \sin 377t)}{dt} \right)$$
$$= (500 \text{ turns})(5 \times 10^{-3} \text{ Wb}) \cos 377t \left(377 \, \frac{\text{rad}}{\text{s}} \right)$$
$$= 942.5 \text{ V} \cos 377t \quad (940 \text{ V} \cos 377t)$$

The answer is D.

Problem 18

A wye-connected, three-phase, 50 Hz, 12-pole synchronous alternator develops an effective (rms) line voltage of 1200 V when the rotor currrent is a steady 4 A. This alternator is rewired to generate 60 Hz voltages. What is the new synchronous speed?

(A) 500 rpm
(B) 600 rpm
(C) 1000 rpm
(D) 1200 rpm

Solution

$$p = 12; f = 60 \text{ Hz}$$

The synchronous speed is

$$n_s = (120) \left(\frac{f}{p} \right) = (120) \left(\frac{60 \text{ Hz}}{12} \right)$$
$$= 600 \text{ rpm}$$

Notice that the "120" term incorporates the conversion from rev/s to rpm.

The answer is B.

Problem 19

A manufacturing plant operates with a large number of induction motors. As a result, the plant has a low power factor and is charged more for power. Costs can be reduced by

(A) decreasing the plant's operating voltage
(B) increasing the plant's resistive load
(C) adding lagging reactive load
(D) adding leading reactive load

Solution

Decreasing the operating voltage will not change the power factor. Increasing the plant's resistive load will improve the power factor, but will also increase the overall power consumption (and, thus, the power bill as well). Adding lagging reactive load will make the power factor worse since the induction motors are already causing a lagging power factor. Thus, the correct choice is adding leading reactive load, such as a power factor correction capacitor.

The answer is D.

ELECTROMAGNETICS

Problem 20

A coil with an inductance of 11 mH has a current of 5 mA passing through it. The magnetic energy stored in the coil is most nearly

(A) 0.14 μJ
(B) 0.55 μJ
(C) 1.6 μJ
(D) 5.6 μJ

Solution

The magnetic energy, U, stored in a conductor whose inductance is L and current is I is

$$U = \tfrac{1}{2} L I^2$$
$$= \left(\tfrac{1}{2} \right) (11 \times 10^{-3} \text{ H})(5 \times 10^{-3} \text{ A})^2$$
$$= 1.375 \times 10^{-7} \text{ J} \quad (0.14 \, \mu\text{J})$$

The answer is A.

Problem 21

A toroidal core has a mean radius of 0.03 m and a cross-sectional area of 1 cm^2. It has a coil of 100 turns wound on it through which passes a current of 3 A. The magnetic field strength, H, is 1591.5 A/m. The core material has a relative permeability of $\mu_r = 2$; the permeability of the free space is $\mu_0 = 4\pi \times 10^{-7}$ H/m. The magnetic flux set up in the toroidal core is most nearly

(A) 0.2 μWb
(B) 0.4 μWb
(C) 0.9 μWb
(D) 1 μWb

Solution
The magnetic flux density is

$$B = \mu_r \mu_0 H$$
$$= (2)\left(4\pi \times 10^{-7} \, \frac{\text{H}}{\text{m}}\right)\left(1591.5 \, \frac{\text{A}}{\text{m}}\right)$$
$$= 0.004 \text{ T}$$
$$A = \frac{1 \text{ cm}^2}{\left(100 \, \frac{\text{cm}}{\text{m}}\right)^2}$$
$$= 1 \times 10^{-4} \text{ m}^2$$
$$\phi = BA = (0.004 \text{ T})(1 \times 10^{-4} \text{ m}^2)$$
$$= 0.4 \times 10^{-6} \text{ Wb} \quad (0.4 \, \mu\text{Wb})$$

The answer is B.

Problem 22
A particle with a charge of $Q = 1.6 \times 10^{-13}$ C is to be moved between points A and B. The points have a potential difference between them of $V_{AB} = 3000$ V. The work required to move the charge is

(A) 4.8×10^{-13} J
(B) 4.8×10^{-10} J
(C) 16×10^{-10} J
(D) 20×10^{-10} J

Solution
The work, W, required is given by

$$W = QV_{AB}$$
$$= (1.6 \times 10^{-13} \text{ C})(3000 \text{ V})$$
$$= 4.8 \times 10^{-10} \text{ J}$$

The answer is B.

Problem 23
A 1 m long conductor is moving steadily with a velocity of 300 m/s in a magnetic field with a flux density, B, of 0.05 T. If the direction of movement is perpendicular to the magnetic flux, the voltage induced in the conductor is

(A) 15 V
(B) 50 V
(C) 100 V
(D) 200 V

Solution
The induced voltage, V, is given by

$$V = BLv = (0.05 \text{ T})(1 \text{ m})\left(300 \, \frac{\text{m}}{\text{s}}\right)$$
$$= 15 \text{ V}$$

The answer is A.

CONTROL SYSTEMS

Problem 24
An open-loop function $f(s)$ is given as

$$f(s) = \frac{K}{s(s+1)(s+4)}$$

What is the location of the poles of $f(s)$ on the real portion of the s-axis?

(A) $0, -1, -4$ rad/s
(B) $-1, -4$ rad/s
(C) $0, 1, 4$ rad/s
(D) 0

Solution
Since there are no s terms in the numerator, there are no zeros. To find the poles, set each denominator term to zero separately.

$$s = 0$$
$$s + 1 = 0$$
$$s = -1$$
$$s + 4 = 0$$
$$s = -4$$

The poles are located at $0, -1$, and -4 rad/s.

The answer is A.

Problem 25
For the control system shown, the position error coefficient, ε_p, is

$$\varepsilon_p = \lim_{s \to 0} \frac{1}{1 + G(s)}$$

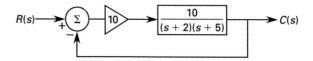

What is the value of the error coefficient for this control system?

(A) 1/110
(B) 1/11
(C) 1/10
(D) 10

Solution
The open-loop gain, $G(s)$, is

$$G(s) = \frac{(10)(10)}{(s+2)(s+5)}$$
$$= \frac{100}{(s+2)(s+5)}$$

The error coefficient is

$$\varepsilon_p = \lim_{s \to 0} \frac{1}{1+G(s)} = \lim_{s \to 0} \frac{1}{1 + \frac{100}{(s+2)(s+5)}}$$

$$= \frac{1}{1 + \frac{100}{10}}$$

$$= 1/11$$

The answer is B.

Problem 26
What is the steady-state gain for the network shown in the following illustration?

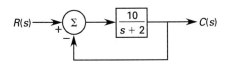

(A) 1/5
(B) 5/6
(C) 6/5
(D) 5

Solution

$$G_R(s) = 1$$
$$G_C(s) = \frac{10}{s+2}$$
$$G_1(s) = 1$$
$$L(s) = 0$$
$$G_2(s) = 1$$

Substituting into the overall system gain equation,

$$G(s) = \frac{C(s)}{R(s)} = \frac{\frac{10}{s+2}}{1 + \frac{10}{s+2}}$$

$$= \frac{10}{s+2+10}$$

$$= \frac{10}{s+12}$$

The steady-state gain is given by

$$G_{ss} = \lim_{s \to 0} G(s) = \frac{10}{0+12}$$
$$= 5/6$$

The answer is B.

Problem 27
What are the roots of the characteristic equation for the system shown?

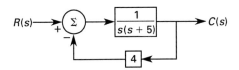

(A) −4
(B) −5
(C) 0, −5
(D) −1, −4

Solution
The characteristic equation is given by

$$G_C(s) = \frac{1}{s(s+5)}$$
$$G_1(s) = 1; G_2(s) = 1; H(s) = 4$$
$$G_C(s)G_1(s)G_2(s)H(s) + 1 = 0$$
$$\frac{4}{s(s+5)} + 1 = 0$$

Multiply both sides by $s(s+5)$ and rearrange.

$$s^2 + 5s + 4 = 0$$

The equation factors into

$$(s+1)(s+4) = 0$$

The roots are $s = -1$ and $s = -4$.

The answer is D.

Problem 28
For a control system whose open-loop gain is $G(s) = 20/(s+10^3)$, what is the approximate break (corner) frequency?

(A) 8.0 Hz
(B) 50 Hz
(C) 160 Hz
(D) 1000 Hz

Solution
The break frequency is the magnitude of that value of s in the denominator of $G(s)$ that causes the denominator to become zero. Set the denominator to zero.

$$s + 10^3 = 0$$
$$s = -10^3$$
$$|s| = 10^3 \text{ rad/s}$$

$$f = \frac{\omega}{2\pi} = \frac{10^3 \, \frac{\text{rad}}{\text{s}}}{2\pi}$$

$$= 159.2 \text{ Hz} \quad (160 \text{ Hz})$$

The answer is C.

COMMUNICATIONS

Problem 29
The Fourier transform of an impulse $a\delta(t)$ of magnitude a is equal to

(A) a
(B) $a \sin t$
(C) $ae^{j2\pi ft}$
(D) $1/a$

Solution
The Fourier transform X of a given signal $x(t)$ is given by

$$X(f) = \int x(t)e^{-j2\pi ft}\,dt$$

Since $\delta(t) = 1$ for $t = 0$ s and is 0 elsewhere, then for $x(t) = \delta(t)$, $X(f) = a$.

The answer is A.

Problem 30
The Fourier transform of a discrete time signal is

(A) a periodic function
(B) a nonperiodic analog function
(C) a nonperiodic discrete function
(D) periodic if the signal is periodic

Solution
The Fourier transform of a discrete time signal is always a periodic function.

The answer is A.

Problem 31
In an AM signal, a 900 kHz carrier is modulated by a music signal that has frequency components from 1 kHz to 10 kHz. The range of the frequencies generated for the upper sidebands is

(A) 440 kHz to 453 kHz
(B) 890 kHz to 899 kHz
(C) 899 kHz to 910 kHz
(D) 901 kHz to 910 kHz

Solution
The highest frequency of the modulating signal is $f_h = 10$ kHz. The lowest frequency of the modulating signal is $f_l = 1$ kHz. The carrier frequency is $f_c = 900$ kHz. The upper sidebands will then include the frequencies from $f_c + f_l$ to $f_c + f_h$. Hence, the range of the upper sidebands is from 901 kHz to 910 kHz.

The answer is D.

Problem 32
An AM radio station is broadcasting at 30 kW and 85% modulation. The power of the sidebands is most nearly

(A) 8 kW
(B) 10 kW
(C) 20 kW
(D) 30 kW

Solution
The equation for the total transmitted power, P_t, is

$$P_t = P_c\left(1 + \frac{m^2}{2}\right)$$

P_t is the total transmitted power of the sidebands and carrier, P_c is the carrier power, and m is the modulation amplitude, which is equal to 0.85. Also, $P_t = P_c + P_s$, where P_s is the power of the sidebands. Therefore,

$$30 \text{ kW} = P_c\left(1 + \frac{m^2}{2}\right) = P_c\left(1 + \frac{(0.85)^2}{2}\right)$$
$$= P_c(1.36)$$

$$P_c = \frac{30 \text{ kW}}{1.36}$$
$$= 22 \text{ kW}$$

$$P_s = P_t - P_c = 30 \text{ kW} - 22 \text{ kW}$$
$$= 8 \text{ kW}$$

The answer is A.

SIGNAL PROCESSING

Problem 33

The analog filter shown has an output $y(t)$ that is sampled at a frequency of 10 Hz, to provide the sequence $y(nT)$. If $x(t) = 0$ for $t < 0$ s, and $x(t) = 1$ for $t \geq 0$ s, then for $n = \{0, 1, 2, 3, ...\}$, $y(nT)$ is given by

(A) $\{0, 0.393, 0.632, 0.777, ...\}$
(B) $\{0, 1.965, 3.16, 3.89, ...\}$
(C) $\{0.393, 0.632, 0.777, ...\}$
(D) $\{1.965, 3.16, 3.89, 4.32, ...\}$

Solution
The input $x(t)$ is a unit step function that has the frequency representation $x(s) = 1/s$.

$$y(s) = \frac{1}{s} \cdot \frac{5}{s+5} = \frac{A}{s} + \frac{B}{s+5}$$
$$= \frac{1}{s} - \frac{1}{s+5}$$

$y(t)$ is the inverse Laplace transform of $y(s)$.

$$y(t) = \mathcal{L}^{-1}\left(\frac{1}{s} - \frac{1}{s+5}\right)$$
$$= \mathcal{L}^{-1}\left(\frac{1}{s}\right) - \mathcal{L}^{-1}\left(\frac{1}{s+5}\right)$$
$$= 1 - e^{-5t}$$

Since $f_s = 10$ Hz, the sampling period is

$$T = \frac{1}{f_s} = \frac{1}{10 \text{ Hz}} = 0.1 \text{ s}$$

The values of $y(nT)$ can be calculated.

$$y(nT)|_{n=0} = y(0) = 1 - e^{-0}$$
$$= 0$$
$$y(nT)|_{n=1} = y(0.1) = 1 - e^{-0.5}$$
$$= 0.393$$
$$y(nT)|_{n=2} = y(0.2) = 1 - e^{-1}$$
$$= 0.632$$
$$y(nT)|_{n=3} = y(0.3) = 1 - e^{-1.5}$$
$$= 0.777$$
$$y(nT) = \{0, 0.393, 0.632, 0.777, ...\}$$

The answer is A.

Problem 34

$x(n) \longrightarrow \boxed{\dfrac{1}{1 - 0.2z^{-1} - 0.1z^{-2}}} \longrightarrow y(n)$

The filter shown represents a

(A) FIR filter
(B) IIR filter
(C) first order digital filter
(D) non-causal filter

Solution
If $h(n)$ is the impulse response of the filter, then

$$X(z)H(z) = Y(z)$$
$$X(z)\left(\frac{1}{1 - 0.2z^{-1} - 0.1z^{-2}}\right) = Y(z)$$
$$X(z) = Y(z)\left(1 - 0.2z^{-1} - 0.1z^{-2}\right)$$

Since the inverse Z-transform of $Z^{-k}Y(z)$ is $y(n-k)$ (shift property), the inverse transform is given by

$$x(n) = y(n) - 0.2y(n-1) - 0.1y(n-2)$$
$$y(n) = x(n) + 0.2y(n-1) + 0.1y(n-2)$$

The impulse response can be determined by letting $x(n) = \delta(n)$

n	$x(n)$	$Y(n) = h(n)$
<0	0	0
0	1	1
1	0	0.2
2	0	0.14
3	0	0.048
4	0	0.024
5	0	0.0096
$\vdots$	$\vdots$	$\vdots$

Although $h(n)$ gets very small as n gets large, it never reaches zero. The impulse response is infinite, and the filter is an infinite impulse response (IIR) filter.

The answer is B.

Problem 35

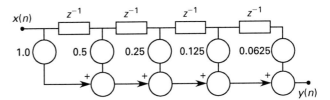

If $x(n) = 120\sin(3\pi n/2)$, $y(2)$ is most nearly

(A) -45
(B) 0
(C) 45
(D) 60

Solution
This is a fourth order FIR filter and therefore depends only on the current value of x and its previous four values. Thus, $x(2)$, $x(1)$, $x(0)$, $x(-1)$ and $x(-2)$ are needed to compute $y(2)$.

$$x(2) = 120\sin(3\pi) = 0$$
$$x(1) = 120\sin\left(\frac{3\pi}{2}\right) = -120$$
$$x(0) = 120\sin(0) = 0$$
$$x(-1) = 120\sin\left(\frac{-3\pi}{2}\right) = 120$$
$$x(-2) = 120\sin(-3\pi) = 0$$
$$y(n) = x(n) + 0.5x(n-1) + 0.25x(n-2)$$
$$\qquad + 0.125x(n-3) + 0.0625x(n-4)$$
$$y(2) = x(2) + 0.5x(2-1) + 0.25x(2-2)$$
$$\qquad + 0.125x(2-3) + 0.0625x(2-4)$$
$$= 0 - 60 + 0 + 15 + 0 = -45$$

The answer is A.

Problem 36
What is the minimum sampling frequency that can be used to avoid aliasing with the following analog signal?

$$x(t) = \cos 100\pi t + \sin 200\pi t + \sin 60\pi t$$

(A) 30 Hz
(B) 60 Hz
(C) 200 Hz
(D) 400 Hz

Solution
To avoid aliasing, the analog signal must be sampled at least at the Nyquist rate, which must be at least twice the highest frequency contained in the signal. The highest frequency contained in the given signal is 200π rad/s or 100 Hz. Therefore, the minimum frequency that can be used to avoid aliasing is 200 Hz.

The answer is C.

Problem 37
The following signal is sampled with a sampling frequency of $f_s = 30$ Hz.

$$x(t) = 5\sin 20\pi t + 4\sin 30\pi t + 3\sin 40\pi t$$

What is the resulting digital signal, $x(n)$?

(A) $x(n) = 5\sin\frac{2}{3}\pi n + 4\sin\frac{1}{2}\pi n + 3\sin\frac{3}{4}\pi n$
(B) $x(n) = 5\sin\frac{2}{3}\pi n + 3\sin\frac{4}{3}\pi n$
(C) $x(n) = 5\sin\frac{2}{3}\pi n + 4\cos\pi n + 3\sin\frac{4}{3}\pi n$
(D) $x(n) = 5\sin\frac{2}{3}\pi n + 4\cos 0.4\pi n + 3\sin\frac{4}{3}\pi n$

Solution
The digital signal is obtained by replacing t with nT where T is the sampling period equal to $1/f_s$. Replacing t with $n/30$,

$$x(n) = 5\sin\tfrac{2}{3}\pi n + 4\sin\pi n + 3\sin\tfrac{4}{3}\pi n$$

Since $\sin\pi n = 0$ for all integer values of n,

$$x(n) = 5\sin\tfrac{2}{3}\pi n + 3\sin\tfrac{4}{3}\pi n$$

The answer is B.

Problem 38

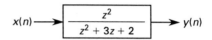

The digital filter shown can be described as

(A) stable
(B) marginally stable
(C) unstable
(D) non-causal

Solution
Convert from filter form to transfer function form by multiplying the numerator and denominator by z^2.

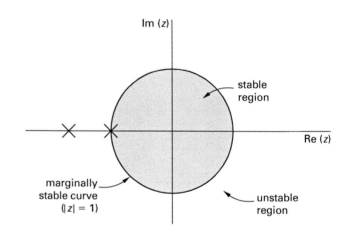

There are two zeros at $z = 0$, and two poles at

$$z = \frac{-3 \pm \sqrt{9-8}}{2} = \frac{-3 \pm (-1)}{2}$$
$$= -2, -1$$

The filter has a pole on the marginally stable curve. It also has one in the unstable region, and therefore is unstable.

The answer is C.

Problem 39
A digital system is known to be unstable with an impulse response given by $h(n) = 1.25^n$. The following circuit is proposed.

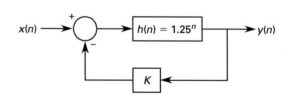

What is the minimum value of K that stabilizes this filter?

(A) $K \geq 0.25$
(B) $K > 0.25$
(C) $K \geq 1.25$
(D) $K > 1.25$

Solution

The closed loop filter has a z-domain transfer function given by

$$\frac{Y(z)}{X(z)} = \frac{H(z)}{1 + KH(z)} = \frac{\frac{1}{1-1.25z^{-1}}}{1 + \frac{K}{1-1.25z^{-1}}}$$

$$= \frac{\frac{1}{1-1.25z^{-1}}}{\frac{1-1.25z^{-1}+K}{1-1.25z^{-1}}}$$

$$= \frac{1}{(1+K) - 1.25z^{-1}}$$

$$= \frac{\frac{1}{1+K}}{1 - \left(\frac{1.25}{1+K}\right)z^{-1}}$$

In order for this filter to be stable,

$$\frac{1.25}{1+K} < 1$$

Rearranging,

$$1.25 < 1 + K$$
$$K > 0.25$$

The answer is B.

ELECTRONICS

Problem 40

A voltmeter is used to measure the voltage across a resistor. The voltmeter is connected in parallel with the resistor. To obtain an accurate voltage reading, the internal resistance of the voltmeter must be

(A) much smaller than the resistance whose voltage is being measured
(B) equal to the resistance whose voltage is being measured
(C) equal to twice the resistance whose voltage is being measured
(D) much greater than the resistance whose voltage is being measured

Solution

Because the voltmeter is connected in parallel with the load resistance, its internal resistance must be very high so that it will not draw any significant current that would affect the original circuit.

The answer is D.

Problem 41

A meter movement has an internal resistance, R_m, of 5 kΩ and full-scale current, I_{fs}, of 50 μA. The meter is to be used to design a voltmeter with a full-scale voltage, V_{fs}, of 20 V. The value of the resistance R_{series} that must be connected in series with the meter to achieve the design goal most nearly is

(A) 2 kΩ
(B) 5 kΩ
(C) 400 kΩ
(D) 500 kΩ

Solution

The maximum voltage drop across the meter is V_{fs}. The maximum current through the meter is 50 μA. Therefore, the total internal resistance of the voltmeter is

$$R_t = \frac{V_{fs}}{I_{fs}} = \frac{20 \text{ V}}{50 \times 10^{-6} \text{ A}}$$
$$= 4 \times 10^5 \; \Omega \quad (400 \text{ k}\Omega)$$

The total resistance is

$$R_t = R_m + R_{series}$$

Hence,

$$R_{series} = R_t - R_m$$
$$= 400 \text{ k}\Omega - 5 \text{ k}\Omega$$
$$= 395 \text{ k}\Omega \quad (400 \text{ k}\Omega)$$

The answer is C.

Problem 42

A sine wave generator is connected to the vertical input of an oscilloscope. The oscilloscope screen has a horizontal width of 10 cm and a vertical height of 8 cm. If the sweep speed of the oscilloscope is set to 4 ms/cm and the number of cycles on the screen is 4, the frequency of the sine wave is most nearly

(A) 60 Hz
(B) 100 Hz
(C) 200 Hz
(D) 500 Hz

Solution

The frequency of the wave is

$$f = \left(\frac{4 \text{ cycles}}{10 \text{ cm}}\right)\left(\frac{1}{4 \frac{\text{ms}}{\text{cm}}}\right)\left(1000 \frac{\text{ms}}{\text{s}}\right)$$

$$= 100 \text{ Hz}$$

The answer is B.

Problem 43

The intrinsic carrier concentration of silicon at room temperature is $n_i = 1.18 \times 10^{10}$ cm^{-3}. A silicon crystal has a hole concentration of $p = 3 \times 10^2$ cm^{-3}. The electron concentration, n, in the crystal is

(A) 1.2×10^{10} cm^{-3}
(B) 2.0×10^{10} cm^{-3}
(C) 4.0×10^{10} cm^{-3}
(D) 4.6×10^{17} cm^{-3}

Solution

Use the mass-action law.

$$np = n_i^2$$

$$n = \frac{n_i^2}{p} = \frac{\left(1.18 \times 10^{10} \frac{1}{\text{cm}^3}\right)^2}{3 \times 10^2 \frac{1}{\text{cm}^3}}$$

$$= 4.6 \times 10^{17} \text{ cm}^{-3}$$

The answer is D.

Problem 44

In metallic materials, electric current occurs due to the movement of electrons. In a semiconductor, there

(A) is one type of current flow
(B) are two types of current flow
(C) are three types of current flow
(D) are four types of current flow

Solution

In a semiconductor there are two types of current flow: electrons flow in one direction, and holes flow in the other direction.

The answer is B.

Problem 45

A silicon crystal is 5 mm long and has a rectangular cross section 20 μm × 40 μm. The donor concentration, N_d, at room temperature is 5×10^{15} cm^{-3}. A current of 2 μA exists in the crystal. If the mobility, μ_n, of an electron is 1.5×10^3 cm^2/V·s and the charge of a hole is $Q = 1.6 \times 10^{-19}$ C, the voltage, V, across the crystal is most nearly

(A) 0.1 V
(B) 0.2 V
(C) 0.3 V
(D) 0.5 V

Solution

$$A = (20 \times 10^{-6} \text{ m})(40 \times 10^{-6} \text{ m})$$
$$= 8 \times 10^{-10} \text{ m}^2 \quad (8 \times 10^{-6} \text{ cm}^2)$$

The conductivity of the crystal is found by

$$\sigma \approx QN_d\mu_n$$
$$= (1.6 \times 10^{-19} \text{ C})\left(5 \times 10^{15} \frac{1}{\text{cm}^3}\right)$$
$$\times \left(1.5 \times 10^3 \frac{\text{cm}^2}{\text{V·s}}\right)$$
$$= 1.2 \text{ 1/}\Omega\text{·cm}$$

$$V = \frac{IL}{A\sigma} = \frac{(2 \times 10^{-6} \text{ A})(0.5 \text{ cm})}{(8 \times 10^{-6} \text{ cm}^2)\left(1.2 \frac{1}{\Omega\text{·cm}}\right)}$$

$$= 0.104 \text{ V} \quad (0.1 \text{ V})$$

The answer is A.

Problem 46

For $t > 0$ s, the output of an operational amplifier is

$$v_o(t) = 10 - 10e^{\frac{-t}{200 \times 10^{-9} \text{ s}}}$$

This output is connected to a logic gate that requires a 4.8 V input to identify its input as a digital "1". What is the approximate time required for the operational amplifier output to reach 4.8 V?

(A) 26 ns
(B) 130 ns
(C) 260 ns
(D) 650 ns

Solution

Compute the time required to reach 4.8 V by replacing $v_o(t)$ with 4.8 V.

$$4.8 \text{ V} = 10 - 10e^{\frac{-t}{200 \times 10^{-9} \text{ s}}}$$

$$-5.2 \text{ V} = -10e^{\frac{-t}{200 \times 10^{-9} \text{ s}}}$$

$$0.52 = e^{\frac{-t}{200\times 10^{-9}\,\text{s}}}$$
$$\ln 0.52 = \ln e^{\frac{-t}{200\times 10^{-9}\,\text{s}}}$$
$$-0.65393 = \frac{-t}{200\times 10^{-9}\,\text{s}}$$
$$t = 1.308 \times 10^{-7}\,\text{s}\quad(130\,\text{ns})$$

The answer is B.

DIGITAL SYSTEMS

Problem 47

A JK flip-flop is subjected to the input of the waveforms shown. The initial state of the flip-flop is 1. Determine the state of the flip-flop after each of the four clock pulses (from left to right).

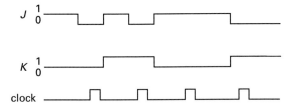

(A) 1-0-1-0
(B) 0-0-0-0
(C) 0-1-0-1
(D) 1-1-1-1

Solution
During the first clock pulse, the inputs are $J = 0$ and $K = 0$. With the initial state 1, the next state is 1. During the second clock pulse, the inputs are $J = 0$ and $K = 1$. The next state is 0. During the third clock pulse, the inputs are $J = 1$ and $K = 0$. The next state is 1. During the fourth clock pulse, the inputs are $J = 0$ and $K = 1$. The next state is 0. So, the sequence is 1-0-1-0.

The answer is A.

Problem 48
What is the rise time for the waveform shown?

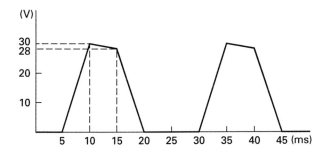

(A) 4 ms
(B) 5 ms
(C) 10 ms
(D) 15 ms

Solution
The rise time of a waveform is the difference between the times to reach the 10% point of its leading edge and the 90% point of that same leading edge. The leading edge of the waveform starts at 5 ms and reaches its peak at 10 ms. Its duration is

$$10\,\text{ms} - 5\,\text{ms} = 5\,\text{ms}$$

For the leading edge of the waveform, the 10% point is 3 V and the 90% point is 27 V.

$$27\,\text{V} - 3\,\text{V} = 24\,\text{V}$$

The rise time is the difference between the times of these two values. Using a direct proportion, the rise time is

$$\tau_{\text{rise}} = (5\,\text{ms})\left(\frac{24\,\text{V}}{30\,\text{V}}\right)$$
$$= 4\,\text{ms}$$

The answer is A.

Problem 49
A logic-gate output Thevenin equivalent circuit is shown. The gate drives wiring and cable capacitance whose total value is 75 pF. What is the effect of the wiring and cable capacitance upon the leading edge of the output waveform?

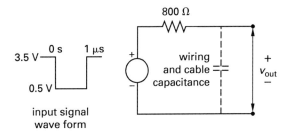

(A) The edge will require 0.03 μs to fall.
(B) The edge will require 0.06 μs to fall.
(C) The edge will require 0.30 μs to fall.
(D) The pulse will be inverted.

Solution
The wiring and cable capacitance time constant is

$$\tau = RC$$
$$= (800\,\Omega)(75 \times 10^{-12}\,\text{F})$$
$$= 6 \times 10^{-8}\,\text{s}\quad(0.06\,\mu\text{s})$$

The effect of the capacitance upon the pulse leading and trailing edges will be essentially negligible in five time constants.

$$5\tau = (5)(0.06 \ \mu s)$$
$$= 0.3 \ \mu s$$

The answer is C.

Problem 50
What is the 4-bit (straight) binary equivalent of the decimal number 13?

(A) 0111
(B) 1011
(C) 1101
(D) 1110

Solution
The decimal number 13 consists of the sum of 8, 4, and 1.

$$13_{10} = 1 \times 2^3 + 1 \times 2^2 + 0 \times 2^1 + 1 \times 2^0$$

Therefore, the binary equivalent is 1101.

The answer is C.

Problem 51
For the digital circuit shown, what are the outputs at Y and Z?

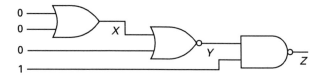

(A) Y is 0, Z is 0
(B) Y is 0, Z is 1
(C) Y is 1, Z is 0
(D) Y is 1, Z is 1

Solution
The first gate from the left is an OR; its output X is "0". The second gate is a NOR; its output Y is "1". The third gate is a NAND; its output Z is "0".

The answer is C.

COMPUTER SYSTEMS

Problem 52
How many memory locations can be directly addressed by a 32-bit microprocessor?

(A) 1 GB
(B) 2 GB
(C) 4 GB
(D) 8 GB

Solution
An n-bit processor can directly address up to 2^n memory locations. Additional locations can only be addressed using indirect schemes.

$$2^n = 2^{32}$$
$$= 4{,}294{,}967{,}296 \quad (4 \text{ GB})$$

The answer is C.

Problem 53
How many inputs can a multiplexer having two selection lines have?

(A) 2
(B) 4
(C) 8
(D) 10

Solution
If the number of selection lines is n, then the number of inputs for the multiplexer is 2^n. With 2 selection lines, 1 input out of 4 can be selected.

The answer is B.

Problem 54
Consider the following circuit.

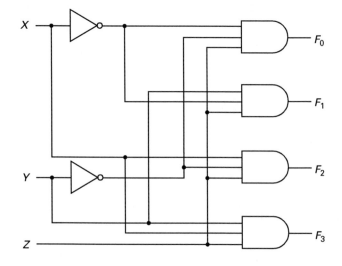

The circuit is a

(A) 2-bit comparator
(B) decoder
(C) full-adder
(D) full-subtractor

Solution

The output F_0 is at logic 1 when the inputs X, Y, Z are at logic 0, 0, 1. The output F_1 is at logic 1 when the inputs X, Y, Z are at logic 0, 1, 1. The output F_2 is at logic 1 when the inputs X, Y, Z are at logic 1, 0, 1. The output F_3 is at logic 1 when the inputs X, Y, Z are at logic 1, 1, 1. Therefore, the circuit is a two-to-four decoder.

The answer is B.

Problem 55

Determine the logical values of the outputs of the following circuit.

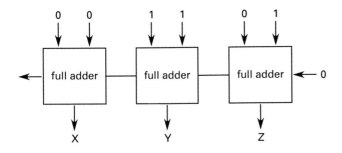

(A) $X = 0, Y = 1, Z = 1$
(B) $X = 1, Y = 1, Z = 0$
(C) $X = 1, Y = 0, Z = 1$
(D) $X = 0, Y = 1, Z = 0$

Solution

The circuit is 2-bit binary adder with the carry-in equal to zero. Thus, $010 + 011 + 000 = 101$ (two + three = five).

The answer is C.

Problem 56

Determine the value of J after the execution of the following program segment.

```
J = 1
FOR I = 1 TO 6 DO
  J = J*I
END
```

(A) 6
(B) 21
(C) 120
(D) 720

Solution

The variable I takes on the values 1, 2, 3, 4, 5, and 6. For each of these values, the computer executes the statement "J = J*I". The variable J takes on the following values 1, 2, 6, 24, 120, 720.

The answer is D.

Problem 57

A queue is a data structure that processes elements in the following order.

(A) first-in, first-out
(B) last-in, first-out
(C) middle-in, last-out
(D) first-in, last-out

Solution

By definition, a queue is a data structure that processes elements first-in, first-out.

The answer is A.

Problem 58

For virtual memory systems, large programs

(A) are executed in memory
(B) can be executed using a smaller memory space
(C) are paged in and out of unused video memory
(D) run faster due to fewer disk accesses

Solution

In a system having virtual memory, large programs are systematically paged into and out of disk space and can be executed using a smaller memory space.

The answer is B.

Problem 59

Which of the following is NOT a programmable register (i.e., a register than can be written to and read) in a typical microprocessor?

(A) ALU
(B) PC
(C) IR
(D) MAR

Solution

PC is a program counter register, used to point to the next instruction in memory. IR is an instruction register, which contains the next instruction fetched from memory. MAR is a memory address register, which points to a location in memory from which data is to be retrieved or to which it is to be stored. All of these are registers that the program can access directly. ALU, however, is the arithmetic logic unit, which performs the basic addition, subtraction, multiplication and division operations required by the program. The ALU is NOT a single register that can be directly accessed by the program.

The answer is A.

Problem 60

How many ASCII characters can be transmitted in 5.0 seconds on a 64 kb communications channel? Assume one bit per character is used for parity checking.

(A) 35 556 characters
(B) 36 409 characters
(C) 40 000 characters
(D) 40 960 characters

Solution

The ASCII code requires a minimum of 7 bits per character, with the most significant bit (MSB) usually set to zero. The code can, however, be used for other purposes, such as in this case where it serves as a parity bit. For that purpose it uses 8 bits per character on a 64 000 bits/s channel.

$$\text{characters} = \left(\frac{64\,000\ \frac{\text{bits}}{\text{s}}}{8\ \frac{\text{bits}}{\text{character}}} \right)(5.0\ \text{sec})$$

$$= 40\,000\ \text{characters}$$

The answer is C.

Practice Exam 1

PROBLEMS

1. A plant engineer is asked to purchase a standby generator to provide backup power to a process fed by a 600 A power distribution panel. The process uses four 100 hp pumps each with an efficiency of 85%. The three-phase motors driving the pumps operate at 480 V and have a power factor of 0.8. What is the minimum size generator that can back up this process in the event of a loss of normal power?

(A) 300 kVA
(B) 350 kVA
(C) 440 kVA
(D) 500 kVA

2. An operational amplifier behaves like a one-pole, low-pass filter. It has a low-frequency gain of 40 dB and a corner frequency of 50 kHz. What is its unity-gain frequency?

(A) 1 MHz
(B) 2 MHz
(C) 5 MHz
(D) 10 MHz

3. What is the desired 3 dB closed-loop (unity gain) bandwidth for a noninverting amplifier if the closed-loop gain is 100 and the unity-gain frequency is 1 MHz?

(A) 1 kHz
(B) 10 kHz
(C) 100 kHz
(D) 1000 kHz

4. An operational amplifier is used in a comparator circuit whose power supply values are ±15 V. The input voltage is a triangular wave that peaks at 20 V on either side of the reference (ground). What type of waveform will appear at the output?

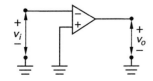

(A) an attenuated square wave
(B) a triangular wave
(C) a square wave
(D) a sine wave

5. A single-pole, low-pass filter is shown. What is its break (corner) frequency?

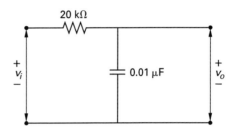

(A) 630 Hz
(B) 800 Hz
(C) 2000 Hz
(D) 5000 Hz

6. A control system has an open-loop gain of 1 (0 dB) at the frequency where its phase angle is −200°. What is the phase margin?

(A) −10°
(B) −20°
(C) −60°
(D) −120°

7. What are the three circuit frequencies for the following circuit?

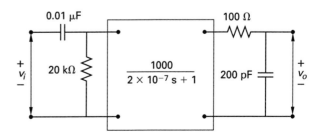

(A) 5 kHz, 5 kHz, and 50 kHz
(B) 800 Hz, 5 MHz, and 8 MHz
(C) 5 kHz, 5 MHz, and 50 kHz
(D) 800 Hz, 800 kHz, and 8 MHz

(A) −3
(B) −2
(C) +2
(D) +3

8. A feedback circuit is constructed as shown. What is the steady-state gain for this network?

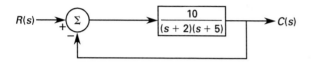

(A) 0.5
(B) 1.0
(C) 10
(D) 20

13. An exponential voltage, v_i, is applied to the circuit shown.
$$v_i = 20e^{\frac{-t}{2\times 10^{-4} \text{ s}}}$$

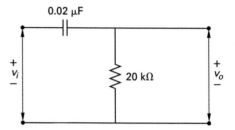

What is the Laplace transform of the output voltage?

(A) $\dfrac{20}{(s+5000)^2}$

(B) $\dfrac{20s}{(s+2500)^2}$

(C) $\dfrac{20}{(s+2500)(s+5000)}$

(D) $\dfrac{20s}{(s+2500)(s+5000)}$

9. A control system has an open-loop gain of 10 dB when its phase angle is −180°. For which of the following gain margins will the system be stable?

(A) −15 dB
(B) −10 dB
(C) −5 dB
(D) +10 dB

10. What is the Nyquist stability criterion for a control system open-loop transfer function?

(A) There must be no zeros in the right-half plane.
(B) The polar plot of $G(s)$ must enclose $-1+j0$.
(C) There must be no poles in the right-half plane.
(D) The polar plot of $G(s)$ must avoid $-1+j0$.

14. What is the pulse repetition frequency for the waveform shown?

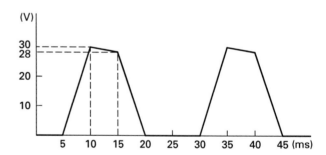

(A) 15 pulses/s
(B) 25 pulses/s
(C) 40 pulses/s
(D) 50 pulses/s

11. The characteristic equation for a control system is $s^2 + 4s + K = 0$. What must be the range of K so that all of the roots will be real?

(A) $K \leq 4$
(B) $K \geq 4$
(C) $K \leq 0$
(D) $K \geq 0$

15. Which of the following is NOT a valid 8-bit word representation for the decimal number −10?

(A) 1 0 0 0 1 0 1 0
(B) 1 1 1 1 0 1 0 1
(C) 1 1 1 1 0 1 1 0
(D) 0 0 0 0 1 0 1 0

12. For an input signal whose frequency is 10^6 rad/s, what will be the slope of a graph of the gain in dB versus the log of frequency for the following open-loop gain equation $G(s)$?

$$G(s) = \frac{50}{(s+10^2)(s+10^4)^2}$$

16. What is the decimal equivalent of the 8-bit binary number 1011 1101?

(A) 76
(B) 85
(C) 153
(D) 189

17. For the digital circuit shown, what are the outputs at Y and Z?

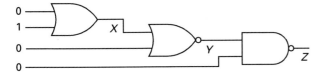

(A) $Y = 0$ and $Z = 0$
(B) $Y = 0$ and $Z = 1$
(C) $Y = 1$ and $Z = 0$
(D) $Y = 1$ and $Z = 1$

18. For the digital circuit shown, what are the outputs at V and X?

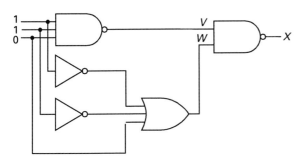

(A) $V = 0$ and $X = 0$
(B) $V = 0$ and $X = 1$
(C) $V = 1$ and $X = 0$
(D) $V = 1$ and $X = 1$

19. For the circuit shown, what is the voltage across the open switch?

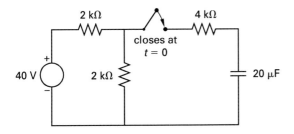

(A) 10 V
(B) 20 V
(C) 30 V
(D) 40 V

20. What is the Thevenin equivalent circuit for the circuit shown?

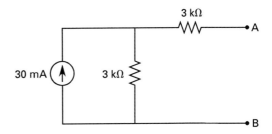

(A) 30 V in series with 1.5 kΩ
(B) 45 V in series with 1.5 kΩ
(C) 90 V in series with 3 kΩ
(D) 90 V in series with 6 kΩ

21. The switch in the following circuit closes at $t = 0$ s. What is the circuit time constant after the switch is closed?

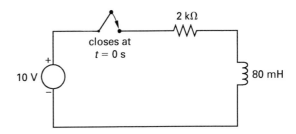

(A) 40 μs
(B) 160 μs
(C) 40 ms
(D) 160 ms

22. Consider the circuit shown.

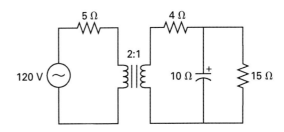

What is the impedance seen by the 120 V source?

(A) $25\,\Omega\angle 0°$
(B) $28\,\Omega\angle 85°$
(C) $45\,\Omega\angle 0°$
(D) $48\,\Omega\angle 35°$

23. What is the voltage across the capacitor once the switch is closed and the following circuit has become stable?

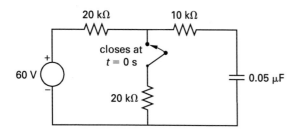

(A) 12 V
(B) 20 V
(C) 30 V
(D) 60 V

24. After the current source has been connected for a long time, the voltage source switches on at $t = 0$ s. What is the voltage across the 4 Ω resistor just prior to the voltage source switching on?

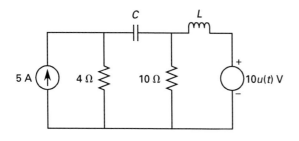

(A) 10 V
(B) 20 V
(C) 25 V
(D) 30 V

25. Two single-phase voltage sources, v_1 and v_2, are connected in series.

$$v_1 = 20 \sin(377t + 25°)$$
$$v_2 = 30 \cos(377t - 15°)$$

What is the total voltage $v_3 = v_1 + v_2$?

(A) $50.0 \tan(377t + 10°)$
(B) $45.5 \sin(377t + 55.3°)$
(C) $10.0 \tan(377t + 10°)$
(D) $45.5 \cos(377t + 55.3°)$

26. A single-phase transformer is rated as 440 V:110 V. If the secondary load is 5.5 Ω resistive and the primary voltage is 440 V, what are the actual primary and secondary currents? Assume an ideal transformer.

(A) 20 A primary, 5 A secondary
(B) 5 A primary, 20 A secondary
(C) 80 A primary, 20 A secondary
(D) 5 A primary, 80 A secondary

27. A wye-connected, three-phase, 50 Hz, 12-pole synchronous alternator develops an effective (rms) line voltage of 1200 V when the rotor currrent is constant at 4 A. This alternator is turned faster to generate 60 Hz voltage. What is the new line voltage for the same rotor current?

(A) 700 V
(B) 1000 V
(C) 1200 V
(D) 1440 V

28. A three-phase delta-connected power supply produces a peak voltage of 170 V at each of its terminals. It is then connected to a purely resistive three-phase load, and the peak line current is measured at 56 A. What is the average power being dissipated in the load?

(A) 3.2 kW
(B) 4.8 kW
(C) 8.3 kW
(D) 9.5 kW

29. Consider the circuit shown.

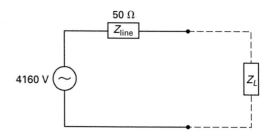

What is most nearly the maximum power that can be delivered to the load Z_L by the 4160 V power source?

(A) 50.0 kW
(B) 87 kW
(C) 170 kW
(D) 350 kW

30. A diode can be formed by doping a pure crystal such that

(A) an n-type semiconductor is produced
(B) it becomes an intrinsic semiconductor
(C) one half of it is p-type and the other half is n-type
(D) a p-type semiconductor is produced

31. A silicon crystal is doped with gallium impurities. The concentration of the impurities is $N_d = 1 \times 10^{10}$ cm^{-3}. The concentration of the free holes, p, in the resulting crystal is approximately

(A) 5×10^5 cm^{-3}
(B) 1×10^{10} cm^{-3}
(C) 3×10^{10} cm^{-3}
(D) 1×10^{20} cm^{-3}

32. A silicon crystal is 5 mm long and has a rectangular cross section of 20 mm × 40 µm. The accepter concentration at room temperature is 5×10^{15} cm^{-3}. If the mobility of the holes, μ_p, is 475 cm^2/V·s and the charge of the holes, Q, is 1.6×10^{-19} C, the conductivity, σ, of the crystal is most nearly

(A) 0.21 1/Ω·cm
(B) 0.38 1/Ω·cm
(C) 3.4 1/Ω·cm
(D) 5.3 1/Ω·cm

33. A bipolar junction transistor usually has

(A) one doped region
(B) two doped regions
(C) three doped regions
(D) four doped regions

34. A linear time-invariant system has a transfer function of $H(s) = 1/(s+3)$. If the system is driven by an impulse signal $5\delta(t)$, the output of the system is

(A) $5e^{-3t}u(t)$
(B) $5 \sin 3\omega t$
(C) $e^{-3\omega t}$
(D) $5 \sin(3\omega t + 30°)$

35. A linear time-invariant system has a transfer function with a direct-current gain equal to 2 and a phase angle of 45°. What is the response of this system to the input $x(t) = 4 \cos 2t$?

(A) $15 \cos 2t$
(B) $8 \cos(2t + 45°)$
(C) $8e^{2t+45°}$
(D) $8 \sin(2t + 45°)$

36. The Fourier transform of a periodic analog signal is

(A) a discrete signal
(B) a continuous time signal
(C) the sum of a discrete and an analog signal
(D) an analog signal equal to the given signal

37. The process of converting the contents of the instruction register into control signals to the various logic and storage elements in the processor is accomplished by

(A) microprogrammed control
(B) pseudocode
(C) instruction register decoding
(D) pipelining

38. A discrete linear time invariant system has an impulse response given by

$$h(n) = (0.5)^{n-1} \quad [n \geq 1]$$
$$= 0, n < 1$$

The difference equation representation for the system is

(A) $y(n) = 0.5x(n-1)$
(B) $y(n) = x(n) + 0.5y(n)$
(C) $y(n) = x(n-1) + 0.5y(n-1)$
(D) $y(n) = 0.5x(n) + y(n-1)$

39. A sine wave generator is connected to the vertical input of an oscilloscope. The scope screen has a horizontal width of 10 cm and a vertical height of 8 cm. The vertical scale of the oscilloscope is 6 V/cm, and the peak-to-peak amplitude of the wave is 4 cm. The maximum voltage of the wave is most nearly

(A) 8 V
(B) 10 V
(C) 30 V
(D) 50 V

40. A linear-time invariant discrete time filter has an impulse response given by $h(n) = (-0.5)^n$. What is the filter's steady state value in response to a unit step?

(A) 0.00
(B) 0.67
(C) 1.0
(D) 2.0

41. A digital filter is empirically observed to have the following output in response to a unit impulse, $\delta(n)$.

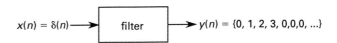

What is the filter's likely response to a unit pulse of three samples (e.g., $x(n) = \{1, 1, 1, 0, 0, 0, ...\}$)?

(A) $\{0, 1, 2, 3, ...\}$
(B) $\{1, 1, 1, 0, ...\}$
(C) $\{0, 1, 3, 6, ...\}$
(D) $\{1, 2, 3, 0, ...\}$

42. A toroidal core has a mean radius of 0.03 m and a cross-sectional area of 1 cm². A coil of 100 turns carrying a current of 3 A is wound on the core. The magnetomotive force, mmf, and the magnetic field intensity, H, produced by the current are most nearly

(A) mmf = 100 A·turns and H = 1000 A·turns/m
(B) mmf = 200 A·turns and H = 1200 A·turns/m
(C) mmf = 300 A·turns and H = 1600 A·turns/m
(D) mmf = 500 A·turns and H = 1900 A·turns/m

43. A charge $Q = 5 \times 10^{-6}$ C is placed in an electric field with intensity $E = 5000$ V/m. The force on the charge exerted by the field is most nearly

(A) 1×10^{-6} N
(B) 25×10^{-3} N
(C) 5×10^{-3} N
(D) 25×10^{-1} N

44. A conductor with a length of 0.4 m carrying a current I is placed in a magnetic field with magnetic flux density of 0.01 T. The field and current direction are orthogonal. If the field exerts a force of 2×10^{-2} N on the conductor, the current in the conductor is most nearly

(A) 1.0 A
(B) 5.0 A
(C) 10 A
(D) 15 A

45. An electric point charge of $Q = 5 \times 10^{-10}$ C is placed in a vacuum. If the permittivity of the free space is 8.854×10^{-12} F/m, the field strength at a distance of 1 μm from the charge is approximately

(A) 1.2×10^{12} V/m
(B) 4.5×10^{12} V/m
(C) 12×10^{12} V/m
(D) 15×10^{12} V/m

46. In a given AM environment, the highest frequency in the modulating signal is 1000 Hz. The frequency of the carrier is 100 kHz. The highest frequency in the AM signal is most nearly

(A) 49 kHz
(B) 51 kHz
(C) 99 kHz
(D) 100 kHz

47. If $x(n)$ is a discrete imaginary and even function, its Fourier transform $X(\omega)$ is

(A) a real and even function
(B) a real and odd function
(C) an imaginary and odd function
(D) an imaginary and even function

48. An analog filter is known to have the following impulse response

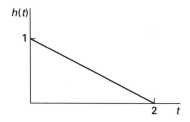

What is the steady-state value of the filter's response to a unit step, $u(t)$?

(A) 0
(B) 1
(C) 2
(D) 3

49. A digital filter has its poles and zeros located in the z-plane as shown.

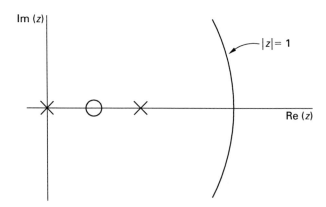

What are the first few terms of the impulse response?

(A) 0, 1.0, 0.25, 0.125, ...
(B) 1.0, 0.25, 0.125, 0.625, ...
(C) 0, 0.25, 0.5, 0.75 ...
(D) 0.25, 0.5, 0.75, 1.0, ...

50. In an AM signal, a 300 W carrier is modulated to the 70% level. The total transmitted power is most nearly

(A) 130 W
(B) 210 W
(C) 300 W
(D) 370 W

51. The bandwidth of a baseband channel is designated for use by

(A) a single analog or digital carrier
(B) two analog or digital carriers
(C) two analog and one digital carriers
(D) two digital carriers

52. What is the value of J after the following program segment is executed?

```
J = 0
WHILE (J < 10) DO
J = J + 3
END
```

(A) 3
(B) 6
(C) 9
(D) 12

53. Which of the following sorting algorithms is the fastest with a truly random list?

(A) bubble sort
(B) insertion sort
(C) quicksort
(D) selection sort

54. In operating systems, what is the meaning of the acronym TLB?

(A) table lookaside buffer
(B) transposed low buffer
(C) temporarily locked buffer
(D) take and look behind buffer

55. What are the following numbers when rounded to three significant digits?

$$X = 3.234$$
$$Y = 1.315$$
$$Z = 1.2349$$

(A) $X = 3.23$, $Y = 1.32$, $Z = 1.23$
(B) $X = 3.23$, $Y = 1.31$, $Z = 1.24$
(C) $X = 3.23$, $Y = 1.32$, $Z = 1.24$
(D) $X = 3.23$, $Y = 1.31$, $Z = 1.23$

56. The process of designing a processor using smaller self-contained building blocks, each with its own ALU, registers, data paths, and control functions is called

(A) integrated circuit design
(B) multi-processor design
(C) bit-slice design
(D) distributed computing

57. The circuit shown depicts a balanced load, where $V_A = 680\sin(377t + 0°)$, $V_B = 680\sin(377t - 120°)$, $V_C = 680\sin(377t - 240°)$, and $Z_1 = Z_2 = Z_3$.

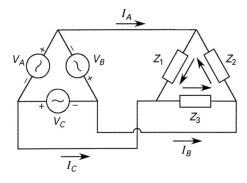

The total power dissipated by the load is 50 kW at a power factor of 0.6 (lagging). The line current I_A is most nearly

(A) $70 \text{ A}\sin(377t + 53°)$
(B) $70 \text{ A}\sin(377t + 23°)$
(C) $140 \text{ A}\sin(377t + 53°)$
(D) $140 \text{ A}\sin(377t + 23°)$

58. What is the function of the circuit shown?

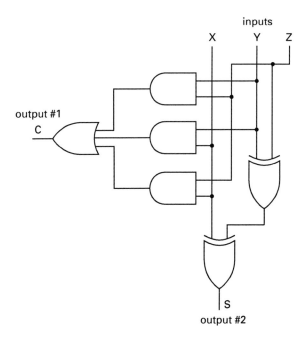

(A) 2-bit comparator
(B) decoder
(C) full-adder
(D) full-subtractor

59. What is the function of the circuit shown?

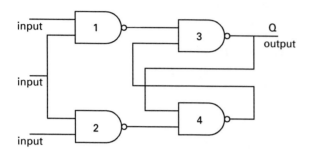

(A) SR flip-flop
(B) JK flip-flop
(C) D-type flip-flop
(D) T-type flip-flop

60. What is the function of the circuit shown?

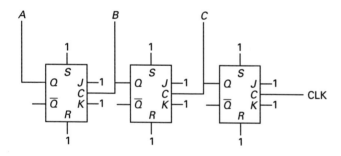

(A) synchronous down-counter
(B) synchronous up-counter
(C) asynchronous down-counter
(D) asynchronous up-counter

SOLUTIONS

1. Calculate the real electrical power needed to provide 100 hp of mechanical power to the process.

$$P = \left(\frac{100 \text{ hp}}{0.85}\right)\left(0.746 \frac{\text{kW}}{\text{hp}}\right)$$
$$= 87.76 \text{ kW}$$

The current drawn by each motor is

$$I_{motor} = \frac{P}{V_{motor}(\text{pf})}$$
$$= \frac{(87.76 \text{ kW})\left(1000 \frac{\text{W}}{\text{kW}}\right)}{(\sqrt{3})(480 \text{ V})(0.8)}$$
$$= 131.9 \text{ A (each motor)}$$

The generator must be capable of supplying 131.9 A at 480 V to each motor. Therefore

$$S_{generator} = \frac{(\sqrt{3})(480 \text{ V})(4)(131.9 \text{ A})}{1000 \frac{\text{VA}}{\text{kVA}}}$$
$$= 438.8 \text{ kVA} \quad (440 \text{ kVA})$$

The answer is C.

2. Start with the gain conversion formula.

$$\text{gain in dB} = 20 \log_{10} K = 20 \log_{10} \frac{v_o}{v_i}$$

Convert the 40 dB gain to a voltage ratio by solving the formula for v_o/v_i.

$$\frac{v_o}{v_i} = \log^{-1} \frac{K}{20} = \log^{-1} \frac{40}{20} = 100$$

The unity-gain frequency is the product of the low-frequency gain (100) and the corner frequency (50 kHz).

$$\text{unity-gain frequency} = (100)(50 \times 10^3 \text{ Hz})$$
$$= 5 \times 10^6 \text{ Hz} \quad (5 \text{ MHz})$$

The answer is C.

3. The closed-loop gain, A_{CL}, is equal to the noise gain, K_n, for a noninverting amplifier. The formula for closed-loop bandwidth, B_{CL}, is

$$B_{CL} = \frac{B}{K_n} = \frac{B}{A_{CL}} = \frac{1 \times 10^6 \text{ Hz}}{100}$$
$$= 10^4 \text{ Hz} \quad (10 \text{ kHz})$$

The answer is B.

4. The output of a comparator circuit switches between two extremes. Therefore, as the triangular input voltage crosses the reference (ground) value, it will cause the output voltage to be either approximately $+15$ V or approximately -15 V, which is less than the input signal voltage. The output is an attenuated square wave.

The answer is A.

5. The break (corner) frequency, ω, in rad/s is the reciprocal of the circuit time constant, τ. The time constant is the product of the R and C values for the filter.

$$\omega = \frac{1}{\tau} = \frac{1}{RC} = \frac{1}{(20 \times 10^3 \ \Omega)(0.01 \times 10^{-6} \ \text{F})}$$
$$= 5 \times 10^3 \ \text{rad/s}$$

$$f = \frac{\omega}{2\pi} = \frac{5 \times 10^3 \ \frac{\text{rad}}{\text{s}}}{2\pi \ \frac{\text{rad}}{\text{cycle}}}$$
$$= 795.8 \ \text{Hz} \quad (800 \ \text{Hz})$$

The answer is B.

6. The phase margin is given by

$$\text{PM} = 180° + \angle G(j\omega_{0 \ \text{dB}})$$
$$= 180° - 200°$$
$$= -20°$$

The answer is B.

7. The break frequencies are the reciprocals of the time constants for each of the two component parts of the circuit. For the black box, the break frequency is the reciprocal of the coefficient of the s term.

$$f_{\text{in}} = \frac{1}{2\pi RC} = \frac{1}{(2\pi)(20 \times 10^3 \ \Omega)(0.01 \times 10^{-6} \ \text{F})}$$
$$= 795.8 \ \text{Hz} \quad (800 \ \text{Hz})$$

$$f_{\text{box}} = \frac{1}{(2\pi)\left(2 \times 10^{-7} \ \frac{\text{s}}{\text{rad}}\right)}$$
$$= 7.958 \times 10^5 \ \text{Hz} \quad (800 \ \text{kHz})$$

$$f_{\text{out}} = \frac{1}{2\pi RC} = \frac{1}{(2\pi)(100 \ \Omega)(200 \times 10^{-12} \ \text{F})}$$
$$= 7.958 \times 10^6 \ \text{Hz} \quad (8 \ \text{MHz})$$

The answer is D.

8.
$$G_R(s) = 1$$
$$G_C(s) = \frac{10}{(s+2)(s+5)}$$
$$G_1(s) = 1; L(s) = 0; G_2(s) = 1$$

Substituting into the overall system gain equation,

$$G(s) = \frac{C(s)}{R(s)} = \frac{\dfrac{10}{(s+2)(s+5)}}{1 + \dfrac{10}{(s+2)(s+5)}}$$
$$= \frac{10}{(s+2)(s+5) + 10}$$
$$= \frac{10}{s^2 + 7s + 20}$$

To determine the steady-state gain, set $s = 0$.

$$\frac{C}{R} = \frac{10}{20} = 0.5 \quad [\text{steady state}]$$

The answer is A.

9. A control system is stable only if its open-loop gain is less than one (0 dB) at the frequency for which the phase angle is $-180°$. The gain margin is expressed in dB. Thus, the gain must be decreased by more than 10 dB at this frequency.

The answer is A.

10. The Nyquist stability criterion may be applied to a control system whose open-loop transfer function has poles or zeros in the right half of the complex plane. For the system to be stable, the polar plot of the open-loop transfer function must not enclose the point $-1 + j0$.

The answer is D.

11. The characteristic equation is

$$s^2 + 4s + K = 0$$

Use the quadratic formula to solve for s.

$$s = \frac{-b \pm \sqrt{b^2 - 4ac}}{2a} = \frac{-4 \pm \sqrt{(4)^2 - (4)(1)K}}{(2)(1)}$$
$$= \frac{-4 \pm \sqrt{16 - 4K}}{2}$$

For the roots to be real,

$$16 - 4K \geq 0$$

Solve for K.

$$K \leq 4$$

The answer is A.

12. For the open-loop gain, $G(s)$, a frequency value of $\omega = 10^6$ rad/s is two decades above the highest break (corner) frequency. The slope of the graph is related to the number of poles in the denominator of $G(s)$. The first pole occurs at -10^2 rad/s; the next poles occur (as a pair) at -10^4 rad/s. The slope of the graph will be the product of the single pole at $\omega = -10^2$ rad/s and the double pole at $\omega = -10^4$ rad/s. The product of the poles is -3.

The answer is A.

13. The output voltage appears across the 20 kΩ resistor. The circuit time constant is

$$\tau = RC = (20 \times 10^3 \ \Omega)(0.02 \times 10^{-6} \ \text{F})$$
$$= 4 \times 10^{-4} \ \text{s}$$

Convert the input to the s-domain (Laplace transform).

$$v_i = \frac{20}{s + \dfrac{1}{200 \times 10^{-6}}} = \frac{20}{s + 5000}$$

The output voltage is the product of the 20 kΩ resistance and the circuit current. In the s-domain, this is

$$v_o = \left(\frac{R}{R + \dfrac{1}{Cs}}\right) v_i = \left(\frac{RCs}{RCs + 1}\right) v_i$$

$$= \left(\frac{4 \times 10^{-4} s}{4 \times 10^{-4} s + 1}\right) v_i$$

$$= \left(\frac{s}{s + 2500}\right) v_i$$

$$= \left(\frac{s}{s + 2500}\right)\left(\frac{20}{s + 5000}\right)$$

$$= \frac{20s}{(s + 2500)(s + 5000)}$$

The answer is D.

14. The pulse repeats every 30 ms $-$ 5 ms $=$ 25 ms. This is its period.

The pulse repetition frequency, PRF, is the reciprocal of the period.

$$\text{PRF} = \frac{1}{25 \times 10^{-3} \ \text{s}} = 40 \ \text{pulses/s}$$

The answer is C.

15. The decimal number 10 requires a minimum of four bits to represent in any notation. When using 8 bits, the most significant bit is always "1" for negative numbers. By inspection answer (D) does not represent -10_{10}. Further examination shows:

1 0 0 0 1 0 1 0 $= -10_{10}$ (signed-binary format)

1 1 1 1 0 1 0 1 $= -10_{10}$ (one's complement format obtained by complementing each bit of the positive binary representation)

1 1 1 1 0 1 1 0 $= -10_{10}$ (two's complement format, obtained by adding 1 to the one's complement format)

0 0 0 0 1 0 1 0 $= +10_{10}$ (same for all three formats)

The answer is D.

16. The decimal values of the ones in each location, from left to right, are 128, 32, 16, 8, 4, and 1. The sum of these six decimal numbers is 189.

The answer is D.

17. The first gate from the left is an OR; its output X is 1. The second gate is a NOR; its output Y is 0. The third gate is a NAND; its output Z is 1.

The answer is B.

18. The first gate is a NAND; its output V is 1. The second gate is an OR; its output W is 0. The third gate is a NAND; its output X is 1.

The answer is D.

19. The voltage across the open switch is the voltage across the vertical 2 kΩ resistor. Use the voltage-divider formula.

$$V_{\text{switch}} = (40 \ \text{V})\left(\frac{2 \times 10^3 \ \Omega}{2 \times 10^3 \ \Omega + 2 \times 10^3 \ \Omega}\right) = 20 \ \text{V}$$

The answer is B.

20. The open-circuit Thevenin equivalent voltage across terminals A and B is the voltage across the vertical 3 kΩ resistor.

$$V_{AB} = RI = (3 \times 10^3 \text{ Ω})(30 \times 10^{-3} \text{ A})$$
$$= 90 \text{ V}$$

The Thevenin equivalent resistance can be determined by reducing the current source to zero. The left leg becomes an open circuit. The resistance across terminals A and B is

$$R_{Th} = 3 \text{ kΩ} + 3 \text{ kΩ}$$
$$= 6 \text{ kΩ}$$

Thus, the Thevenin equivalent circuit across terminals A and B is a series voltage source of 90 V and a resistance of 6 kΩ.

The answer is D.

21. The circuit time constant is

$$\tau = \frac{L}{R} = \frac{80 \times 10^{-3} \text{ H}}{2 \times 10^3 \text{ Ω}}$$
$$= 40 \times 10^{-6} \text{ s} \quad (40 \text{ μs})$$

The answer is A.

22. The transformer secondary impedance is given by

$$Z_{\text{secondary}} = 4\text{ Ω} + (-j10\text{ Ω} \| 15\text{ Ω})$$
$$= 4\text{ Ω} + \frac{150\text{ Ω}\angle -90°}{15 - j10\text{ Ω}}$$
$$= 4\text{ Ω} + \frac{150\text{ Ω}\angle -90°}{18.03\text{ Ω}\angle -33.7°}$$
$$= 4\text{ Ω} + 8.3\angle -56.3°$$
$$= 11.0\text{ Ω}\angle -38.7°$$

Reflecting this impedance to the primary side yields

$$Z_{\text{primary}} = 5\text{ Ω} + (2)^2 (11.0\text{ Ω}\angle -38.7°)$$
$$= 5\text{ Ω} + 44\text{ Ω}\angle -38.7°$$
$$= 5\text{ Ω} + 137.4\text{ Ω} + j110.0\text{ Ω}$$
$$= 39.31\text{ Ω} - j27.55\text{ Ω}$$
$$= 48\text{ Ω}\angle 35°$$

The answer is D.

23. Once the circuit has become stable, the capacitor acts like an open circuit. The 10 kΩ resistor has no current through it. Therefore, the voltage across the capacitor is the same as the voltage across the vertical 20 kΩ resistor. This value of voltage can be calculated from the voltage-divider formula.

$$V_C = (60 \text{ V}) \left(\frac{20 \times 10^3 \text{ Ω}}{20 \times 10^3 \text{ Ω} + 20 \times 10^3 \text{ Ω}} \right)$$
$$= 30 \text{ V}$$

The answer is C.

24. Just prior to $t = 0$ s, the voltage source acts like a short circuit, the capacitor acts like an open circuit, and the inductor acts like a short circuit.

The equivalent circuit at $t = 0^-$ s is

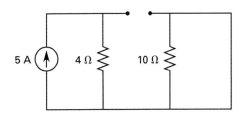

The only path for the current-source current is through the 4 Ω resistor. Applying Ohm's law,

$$V_{4\Omega} = (4 \text{ Ω})(5 \text{ A}) = 20 \text{ V}$$

The answer is B.

25. Express v_2 as a sine function.

$$v_2 = 30 \cos(377t - 15°)$$
$$= 30 \sin(377t + 75°)$$

Convert v_1 and v_2 to phasor form.

$$v_1 = 20(\cos 25° + j \sin 25°)$$
$$= 18.126 + j8.452$$
$$v_2 = 30(\cos 75° + j \sin 75°)$$
$$= 7.765 + j28.978$$

The phasor sum of v_1 and v_2 is

$$v_3 = 25.891 + j37.430$$

Convert v_3 back to the time domain.

$$v_3 = 45.51 \sin(377t + 55.33°)$$

The answer is B.

26. The turns ratio is

$$\frac{N_1}{N_2} = \frac{V_2}{V_1} = \frac{440 \text{ V}}{110 \text{ V}} = 4$$

The secondary current is

$$I_2 = \frac{110 \text{ V}}{5.5 \text{ }\Omega} = 20 \text{ A}$$

The primary current is

$$I_1 = \left(\frac{N_2}{N_1}\right) I_2 = \frac{20 \text{ A}}{4} = 5 \text{ A}$$

The answer is B.

27. The line voltage is equal to the terminal voltage; the terminal voltage is proportional to frequency.

$$\frac{v_{\text{line 2}}}{v_{\text{line 1}}} = \frac{f_2}{f_1}$$

Therefore,

$$v_{\text{line 2}} = v_{\text{line 1}} \left(\frac{f_2}{f_1}\right) = (1200 \text{ V})\left(\frac{60 \text{ Hz}}{50 \text{ Hz}}\right)$$
$$= 1440 \text{ V}$$

The answer is D.

28. The voltages and currents are given in terms of their peak values, and must be converted to their RMS equivalents.

$$V_L = 170 \text{ V}(\sin\omega t + \phi) = \frac{170 \text{ V}\angle 0°}{\sqrt{2}}$$
$$= 120 \text{ V}\angle 0°$$
$$I_L = 56 \text{ A}(\sin\omega t + \phi) = \frac{56 \text{ A}\angle 0°}{\sqrt{2}}$$
$$= 40 \text{ A}\angle 0°$$

Since the load is resistive, the power factor is 1.0. Average power is

$$P = \sqrt{3} V_L I_L \text{pf}$$
$$= \sqrt{3}(120 \text{ V})(40 \text{ A})(1.0)$$
$$= 8.3 \text{ kW}$$

The answer is C.

29. The voltage at the load Z_L is found from a voltage divider calculator.

$$V_L = (4160 \text{ V})\left(\frac{Z_L}{Z_L + 50 \text{ }\Omega}\right)$$

This expression is maximized for large values of Z_L. The current through the load Z_L is given by

$$I_L = \frac{4160 \text{ V}}{Z_L + 50 \text{ }\Omega}$$

The current is maximized for small values of Z_L. The power delivered to the load is given by

$$P_L = \left(\frac{4160 \text{ V}}{Z_L + 50 \text{ }\Omega}\right)^2 (Z_L)$$

Notice that if Z_L is small (i.e., zero), then $P = 0$; and if Z is large (i.e., ∞) then $P = 0$, as well. By inspection, the maximum occurs when $Z_L = 50 \text{ }\Omega$.

$$P_{\max} = \left(\frac{4160 \text{ V}}{50 \text{ }\Omega + 50 \text{ }\Omega}\right)^2 (50 \text{ }\Omega)$$
$$= 86.5 \text{ kW} \quad (87 \text{ kW})$$

The answer is B.

30. A diode is a junction of n-type and p-type semiconductors.

The answer is C.

31. Gallium is a donor-type impurity. Doping silicon with gallium produces a p-type crystal. In a p-type crystal, the concentration of the free holes is approximately equal to the concentration of the donor atoms, as shown by

$$p_p \approx N_a = 1 \times 10^{10} \text{ cm}^{-3}$$

The answer is B.

32. In a p-type crystal, the hole contraction is approximately equal to the acceptor concentration. The conductivity of the crystal is shown by

$$\sigma \approx QN_a\mu_p$$
$$= (1.6 \times 10^{-19} \text{ C})\left(5 \times 10^{15} \frac{1}{\text{cm}^3}\right)\left(475 \frac{\text{cm}^2}{\text{V·s}}\right)$$
$$= 0.38 \text{ 1/}\Omega\text{·cm}$$

The answer is B.

33. A bipolar junction transistor is either a *npn* or a *pnp* device. Therefore, any bipolar transistor has three doped regions.

The answer is C.

34. The response is the inverse Laplace transform of $5\delta(t)/(s+3)$, which is equal to $5e^{-3t}u(t)$.

The answer is A.

35. The response of a linear time-invariant system to a sinusoidal input is (1) a sinusoid with the same frequency as the input, (2) an amplitude equal to the amplitude of the input multiplied by the direct-current gain of the system, and (3) a phase shift equal to the angle of the transfer function of the system.

The amplitude is

$$(4)(\text{direct-current gain}) = (4)(2) = 8$$

The angular frequency is 2. The phase shift is 45°.

The answer is B.

36. The Fourier transform is a series of spikes in the frequency domain only at the repetition frequency and integer multiples of that frequency. The shape of the function determines the amplitude of the spikes. Therefore, the Fourier transform of a periodic analog signal is a discrete function. The Fourier transform of a periodic function in the frequency domain is sometimes called the *Fourier series*, and the amplitude of the spikes are called the *Fourier coefficients*. When a periodic function is expressed as the sum of sine and cosine functions, it is called the *Fourier series expansion*.

The answer is A.

37. *Pseudocode* is a method of writing lines of code that are like English and bear resemblance to a programming language, but will not compile and run. *Pipelining* involves the process of executing parts of one instruction while fetching the data for another. *Instruction register decoding* is an important part of a hard-wire controlled CPU, but does not perform the complete function described. The process that performs the entire function of converting instruction register contents into a complete sequence of control signals is *microprogrammed control*.

The answer is A.

38. Let $h'(n) = 0.5^n$. From the z-transform pair table, $H'(z) = 1/(1 - 0.5z^{-1})$

Since $h(n) = h'(n-1)$, $H(z) = H'(z)z^{-1}$ (shift property).

$$H(z) = \frac{z^{-1}}{1 - 0.5z^{-1}} = \frac{Y(z)}{X(z)}$$

$$Y(z) - 0.5Y(z)z^{-1} = z^{-1}X(z)$$

Taking the inverse z-transform of this expression yields

$$y(n) - 0.5y(n-1) = x(n-1)$$

or

$$y(n) = x(n-1) + 0.5y(n-1)$$

The answer is C.

39. The peak-to-peak value of the wave is

$$V_{pp} = (4 \text{ cm})\left(6 \frac{\text{V}}{\text{cm}}\right) = 24 \text{ V}$$

The maximum value of the wave is

$$V_{max} = \frac{V_{pp}}{2} = \frac{24 \text{ V}}{2} = 12 \text{ V} \quad (10 \text{ V})$$

The answer is B.

40. The filter's Z-domain representation as obtained from the transform pair table is

$$H(z) = \frac{1}{1 + 0.5z^{-1}}$$

If the input is a unit step (i.e., $x(n) = u(n)$) then

$$X(z) = \frac{1}{1 - z^{-1}}$$

$$Y(z) = X(z)H(z)$$

$$= \frac{1}{(1 - z^{-1})(1 + 0.5z^{-1})}$$

Using the final value theorem,

$$\lim_{n \to \infty} y(n) = \lim_{z \to 1}(1 - z^{-1})Y(z)$$

$$= (1 - z^{-1})\left(\frac{1}{(1 - z^{-1})(1 + 0.5z^{-1})}\right)$$

$$= \lim_{z \to 1}\left(\frac{1}{1 + 0.5z^{-1}}\right)$$

$$= \frac{1}{1 + 0.5}$$

$$= 0.667 \quad (0.67)$$

The answer is B.

41. Obtain $y(n)$ by convolution.

$$y(n) = x(n) * f(n)$$
$$= \sum_{k=-\infty}^{\infty} x(k)f(n-k)$$

$$y(0) = \sum_{k=-\infty}^{\infty} x(k)f(-k)$$
$$= x(0)f(0)$$
$$= (1)(0)$$
$$= 0$$

$$y(1) = \sum_{k=-\infty}^{\infty} x(k)f(1-k)$$
$$= x(0)f(1) + x(1)f(0)$$
$$= (1)(1) + (1)(0)$$
$$= 1$$

$$y(2) = \sum_{k=-\infty}^{\infty} x(k)f(2-k)$$
$$= x(0)f(2) + x(1)f(1) + x(2)f(0)$$
$$= (1)(2) + (1)(1) + (1)(0)$$
$$= 3$$

$$y(3) = \sum_{k=-\infty}^{\infty} x(k)f(3-k)$$
$$= x(0)f(3) + x(1)f(2) + x(2)f(1)$$
$$= (1)(3) + (1)(2) + (1)(1)$$
$$= 6$$

The answer is C.

42. The magnetomotive force, mmf, is given by

$$\text{mmf} = NI = (100 \text{ turns})(3 \text{ A}) = 300 \text{ A·turns}$$

The mean path length is

$$L = 2\pi r = (2\pi)(0.03 \text{ m}) = 0.188 \text{ m}$$

The magnetic field intensity, H, is

$$H = \frac{\text{mmf}}{L} = \frac{300 \text{ A·turns}}{0.188 \text{ m}}$$
$$= 1596 \text{ A·turns/m} \quad (1600 \text{ A·turns/m})$$

The answer is C.

43.
$$F = QE$$
$$= (5 \times 10^{-6} \text{ C})\left(5000 \, \frac{\text{V}}{\text{m}}\right)$$
$$= 25 \times 10^{-3} \text{ N}$$

The answer is B.

44. The force exerted by the field on the conductor is the product of the current, length, and magnetic flux density.

$$F = ILB \quad [\text{magnitude}]$$
$$I = \frac{F}{BL} = \frac{2 \times 10^{-2} \text{ N}}{(0.01 \text{ T})(0.4 \text{ m})}$$
$$= 5.0 \text{ A}$$

The answer is B.

45. The electric field strength, E, at a distance r from the charge Q is given by

$$E = \frac{Q}{4\pi\epsilon_0 r^2}$$
$$= \frac{5 \times 10^{-10} \text{ C}}{(4\pi)\left(8.854 \times 10^{-12} \, \frac{\text{F}}{\text{m}}\right)(1 \times 10^{-6} \text{ m})^2}$$
$$= 4.5 \times 10^{12} \text{ V/m}$$

The answer is B.

46. In an AM environment, the frequencies of the modulating signal are shifted by the carrier frequency. The resulting signal will have its frequencies concentrated around the carrier frequency. Therefore, the highest frequency in the AM signal will be 101 kHz (100 kHz).

The answer is D.

47. The Fourier transform of a discrete signal $x(n)$ is given by

$$X(\omega) = \sum x(n)e^{-j\omega n}$$
$$= \sum x(n)(\cos \omega n + j \sin \omega n)$$
$$= \sum x(n) \cos \omega n + j \sum x(n) \sin \omega n$$

The second term is equal to zero since $\sin \omega n$ is an odd function and $x(n)$ is an even function. From the first term, $x(n)$ is an imaginary function, so $X(\omega)$ is an imaginary function. Since $\cos \omega n$ is an even function and $x(n)$ is an even function, $X(\omega)$ is an even function also.

The answer is D.

48. The output $y(t)$ can be obtained most easily by using convolution

$$y(t) = h(t) * u(t)$$
$$= \int_{-\infty}^{\infty} h(\tau) u(t - \tau) \, d\tau$$

Graphically, this is shown by

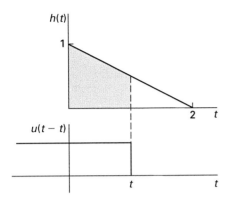

For values of $t < 0$, the integral is zero. For values of t such that $0 \leq t \leq 2$, the integral is the trapezoidal area under the $h(t)$ response curve. For values of $t > 2$, the integral is the complete triangular area under the $h(t)$ response curve.

$$y(t) = \left(\tfrac{1}{2}\right)(1)(2) = 1$$

The answer is B.

49. By inspection,

$$H(z) = \frac{(z - 0.25)}{z(z - 0.5)} = \frac{z - 0.25}{z^2 - 0.5z}$$
$$= \frac{z^{-1} - 0.25z^{-2}}{1 - 0.5z^{-1}}$$
$$= \frac{Y(z)}{X(z)}$$

Use synthetic division.

$$H(z) = 1 - 0.5z^{-1} \overline{)\begin{array}{l} 0 + z^{-1} + 0.25z^{-2} + 0.125z^{-1} \\ z^{-1} - 0.25z^{-2} \\ \underline{z^{-1} - 0.5z^{-2}} \\ 0.25z^{-2} \\ \underline{0.25z^{-2} - 0.125z^{-3}} \\ 0.125z^{-3} \\ \underline{0.125z^{-3} - 0.0625z^{-4}} \\ \vdots \end{array}}$$

$$H(z) = 0 + 1.0z^{-1} + 0.25z^{-2} + 0.125z^{-3} + 0.0625z^{-4}\ldots$$
$$h(n) = \{0, 1.0, 0.25, 0.125, 0.0625, \ldots\}$$

The answer is A.

50. The total transmitted power is

$$P_t = P_c\left(1 + \frac{m^2}{2}\right)$$

P_t is the total transmitted power of the sidebands and carrier, P_c is the carrier power, and m is the modulation index, which in this problem is equal to 0.7.

$$P_t = (300 \text{ W})\left(1 + \frac{(0.7)^2}{2}\right)$$
$$= 373.5 \text{ W} \quad (370 \text{ W})$$

The answer is D.

51. The bandwidth of a baseband channel is designated for use by a single analog or digital carrier.

The answer is A.

52. Since the control will exit the DO loop when J becomes larger than or equal to 10, J has the following values: 0, 3, 6, 9, and 12. When J is 12, the program ends.

The answer is D.

53. Quicksort has the most efficient performance of the given algorithms. It partitions the list into two sublists and reduces the problem to sorting the smaller lists.

The answer is C.

54. A table lookaside buffer (TLB) is used to improve the performance of address translation in virtual memory systems.

The answer is A.

55. When rounding, the least significant retained digit is rounded up if the first discarded digit is larger than four, or if the first discarded digit is a five followed by zeros and the least significant retained digit is odd. Therefore, the value 3.234 is rounded to 3.23, the value 1.315 is rounded to 1.32, and the value 1.2349 is rounded to 1.23.

The answer is A.

56. Integrated circuit design is the process of putting multiple components on a single chip, but is not limited to microprocessors. Multi-processor design involves the use of more than one processor in a single computer. Distributed computing describes the environment where the program and data of a given application are distributed among multiple computers. The answer that most accurately matches the description provided is bit-slice design, where the n-bit data path of a given process is implemented by a m smaller q-bit units where $n = mq$.

The answer is C.

57. Converting the source voltages to phasors yields

$$V_A = \frac{680\,\text{V}\angle 0°}{\sqrt{2}}$$
$$= 480\,\text{V}\angle 0°$$
$$V_B = 480\,\text{V}\angle -120°$$
$$V_C = 480\,\text{V}\angle -240°$$

The power delivered to the load is

$$P = \sqrt{3}V_L I_L \text{pf}$$
$$= (\sqrt{3})(480\,\text{V})I_L 0.6$$
$$= (500\,\text{V})I_L$$
$$I_L = \frac{P}{500\,\text{V}}$$
$$= \frac{50 \times 10^3\,\text{W}}{500\,\text{V}}$$
$$= 100\,\text{A}$$

Since the load is at a power factor, pf, of 0.6, its phase angle is $\theta = \cos^{-1}\text{pf} = 53°$ (lagging). In a balanced delta configuration, the line current leads the phase current by 30°, therefore

$$I_L = 100\,\text{A}\angle 53° - 30°$$
$$= 100\,\text{A}\angle 23°$$
$$= 141\,\text{A}\sin(377t + 23°) \quad (140\,\text{A}\sin(377t + 23°))$$

The answer is D.

58. The circuit has three inputs (X, Y, Z) and two outputs (C, S). The output S is at logic one when an odd number of the inputs is at logic one. The output C is at logic one when any pair of the inputs is at logic one. Therefore, the circuit is a full-adder.

The answer is C.

59. The two output gates (3, 4) form an SR-latch. The two input gates (1, 2) cause the SR-latch to become a clocked flip-flop. The circuit is an SR flip-flop.

The answer is A.

60. Since the clock input of two of the flip-flops is driven by the output of the flip-flops, the counter is asynchronous. Assume the state of the counter is 0, 0, 0. After applying one clock pulse, the next state becomes 1, 1, 1. Therefore, the counter is an asynchronous down-counter.

The answer is C.

Practice Exam 2

PROBLEMS

1. What is the amplitude of v_o for the filter circuit shown?

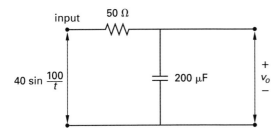

- (A) 0 V
- (B) 20 V
- (C) 28 V
- (D) 40 V

2. The binary number given is in two's complement form and represents a distance in meters.

LSB						MSB
1	0	0	1	1	0	1

LSB is the least significant bit. MSB is the most significant bit. The value of the LSB represents 2.4 m. What is most nearly the distance in base 10?

- (A) 24 m
- (B) 39 m
- (C) 40 m
- (D) 94 m

3.

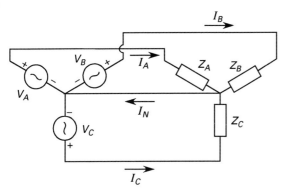

For the circuit shown,

$$V_A = 480 \text{ V} \angle 0°$$
$$V_B = 480 \text{ V} \angle 120°$$
$$V_C = 480 \text{ V} \angle 240°$$
$$Z_A = 50 \, \Omega$$
$$Z_B = 40 + j30 \, \Omega$$
$$Z_C = 43.3 + j25 \, \Omega$$

The neutral current, I_N, is most nearly

- (A) 0 A
- (B) 5.3 A$\angle 63°$
- (C) 5.9 A$\angle 83°$
- (D) 13 A$\angle 87°$

4. The voltage-regulator circuit shown must provide a voltage of 20 V $\pm$ 2 V from an input of 28 V $\pm$ 1 V with a derating of 50% on the resistors. The current to the load is not significant.

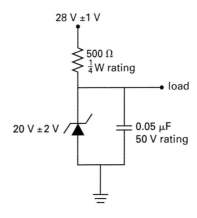

Which of the following component changes can make the circuit meet the requirements?

- (A) Use a 500 Ω resistor with a 0.5 W rating.
- (B) Use a 400 Ω resistor with a 0.25 W rating.
- (C) Use a capacitor with a 20 V rating.
- (D) The circuit meets the requirements as is.

5. What is the binary coded decimal (BCD) equivalent of the following formula?

$$\frac{1101_2 - A_{16} + 34_8 + 17_{10}}{22_3}$$

(A) 0110_{BCD}
(B) 1000_{BCD}
(C) 10100_{BCD}
(D) 110010_{BCD}

6. Consider a control system governed by the gain and phase diagrams shown.

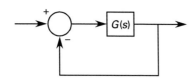

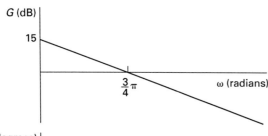

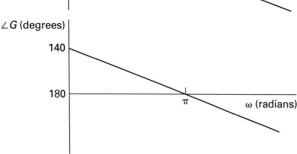

The gain and phase margins are most nearly

(A) gain margin = −10 dB, phase margin = 20°
(B) gain margin = 5 dB, phase margin = 10°
(C) gain margin = 15 dB, phase margin = 22°
(D) gain margin = 20 dB, phase margin = 24°

7. A motor with three pole pairs is operating at 60 Hz. The rotational speed is 1080 rpm. The slip angular velocity is most nearly

(A) π rad/s
(B) $\frac{2\pi}{3}$ rad/s
(C) 2π rad/s
(D) 4π rad/s

8. What is the current through the 2 Ω resistor in the circuit shown?

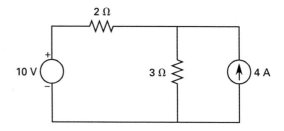

(A) 0.4 A, right to left
(B) 0.4 A, left to right
(C) 1.7 A, right to left
(D) 1.7 A, left to right

9. Given the following equations, determine the time-invariant convolution $h(t) = x(t) * y(t)$.

$$x(t) = \begin{cases} 5 & 0 < t < 1 \\ 0 & \text{elsewhere} \end{cases}$$

$$y(t) = e^{5t}$$

(A) e^{5t}
(B) $e^5 - 1$
(C) $e^{5t}(1 - e^{-5})$
(D) ∞

10. The initial states of the flip-flops in the figure shown are $Q_1 = 0$ and $Q_2 = 0$.

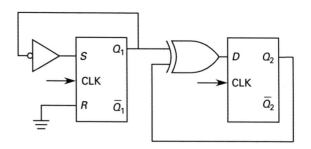

The next state after the clock pulse will be

(A) $Q_1 = 0$, $Q_2 = 0$
(B) $Q_1 = 0$, $Q_2 = 1$
(C) $Q_1 = 1$, $Q_2 = 0$
(D) $Q_1 = 1$, $Q_2 = 1$

11. The truth table and arrangement for the seven-segment display are shown.

inputs A B C D	segments a b c d e f g	decimal number
0 0 0 0	1 1 1 1 1 1 0	0
0 0 0 1	0 1 1 0 0 0 0	1
0 0 1 0	1 1 0 1 1 0 1	2
0 0 1 1	1 1 1 1 0 0 1	3
0 1 0 0	0 1 1 0 0 1 1	4
0 1 0 1	1 0 1 1 0 1 1	5
0 1 1 0	1 0 1 1 1 1 1	6
0 1 1 1	1 1 1 0 0 0 0	7
1 0 0 0	1 1 1 1 1 1 1	8
1 0 0 1	1 1 1 1 0 1 1	9
1 0 1 0	X X X X X X X	not used
1 0 1 1	X X X X X X X	not used
1 1 0 0	X X X X X X X	not used
1 1 0 1	X X X X X X X	not used
1 1 1 0	X X X X X X X	not used
1 1 1 1	X X X X X X X	not used

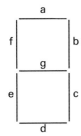

What is the canonical product-of-sum form (POS) of the Boolean formula for segment g?

(A) $(\overline{A}+B+\overline{C}) \cdot (\overline{A}+\overline{B}+\overline{C})$
(B) $(\overline{A}+B+\overline{C}+\overline{D}) \cdot (\overline{A}+\overline{B}+\overline{C}+D)$
(C) $(B+\overline{C}+\overline{D}) \cdot (\overline{A}+\overline{B}+\overline{C})$
(D) $(A+B+C) \cdot (\overline{B}+\overline{C}+\overline{D})$

Problems 12 and 13 are based on the following scenario. The modulation index of a frequency-modulated signal is known to be a sine wave of frequency ω_m. The carrier frequency is known to be ω_c.

12. The phase of the modulated signal has the form

(A) $\omega_c - \dfrac{k_f \cos \omega_m t}{\omega_m t}$
(B) $\omega_c + k_f \sin \omega_m t$
(C) $\omega_c + \omega_m$
(D) $\omega_c \pm \omega_m$

13. The instantaneous frequency of the modulated signal has the form

(A) $\omega_c t - \dfrac{k_f \cos \omega_m t}{\omega_m}$
(B) $\omega_c t + k_f t \sin \omega_m t$
(C) $(\omega_c + \omega_m)t$
(D) $(\omega_c \pm \omega_m)t$

14. A modulation baseband signal with the Laplace transform shown is filtered through an ideal filter with the Laplace gain transform shown.

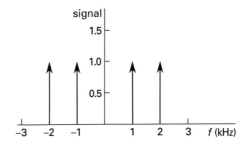

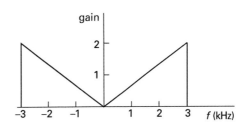

The result is then modulated on a 10 MHz carrier. What is the spectrum of the resulting output?

(A)
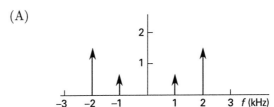

(B)
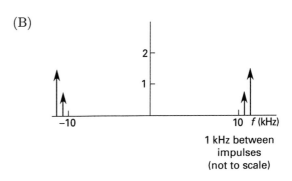

1 kHz between impulses (not to scale)

(C)

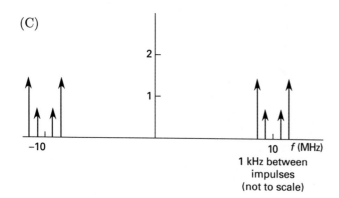

1 kHz between impulses (not to scale)

(D)

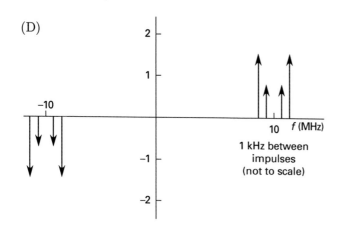

1 kHz between impulses (not to scale)

15. The output of an operational amplifier is $v_o(t) = (10\text{ V})(1 - e^{-t/200 \text{ ns}})$ for $t > 0$ s. What is the approximate amplifier rise time?

(A) 21 ns
(B) 94 ns
(C) 220 ns
(D) 440 ns

16. Consider the circuit shown.

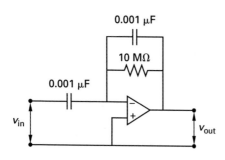

The voltage gain (ignoring phase change) at a 100 Hz input frequency is most nearly

(A) 1.0
(B) 1.5
(C) 2.0
(D) 2.5

17. Consider the circuit shown.

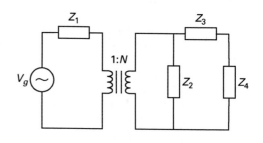

What value of the turns ratio, N, results in the maximum transfer of power from the generator V_g to the secondary side of the transformer?

(A) $N = \dfrac{Z_1(Z_2 + Z_3 + Z_4)}{Z_2(Z_3 + Z_4)}$

(B) $N = \dfrac{Z_2(Z_3 + Z_4)}{Z_1(Z_2 + Z_3 + Z_4)}$

(C) $N = \sqrt{\dfrac{Z_2(Z_3 + Z_4)}{Z_1(Z_2 + Z_3 + Z_4)}}$

(D) $N = \sqrt{\dfrac{Z_1(Z_2 + Z_3 + Z_4)}{Z_2(Z_3 + Z_4)}}$

18. Consider the circuit shown.

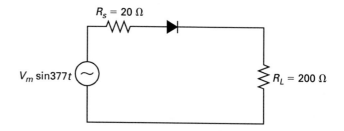

Assume the diode behaves ideally. Most nearly what value of V_m will dissipate 25 W in the load, R_L?

(A) 100 V
(B) 110 V
(C) 140 V
(D) 160 V

19. The number 2 is entered into cell A1 in a spreadsheet. The formula A1 + A1 is entered in cell B1. The contents of cell B1 are then copied and pasted into cell C1. The number is displayed in cell C1 is

(A) 2
(B) 4
(C) 6
(D) 8

20. Which figure represents the Boolean expression $W \oplus X \oplus Y \oplus Z$, where $\oplus$ is the symbol for "exclusive OR"?

(A)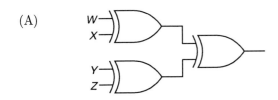

(B)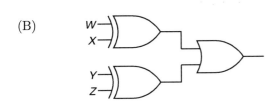

(C)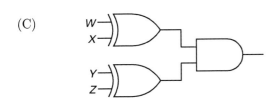

(D)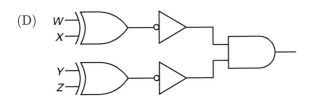

21. The ideal three-phase transformer shown has a primary phase voltage of 1.2 kV and a secondary line-to-line voltage of 208 V.

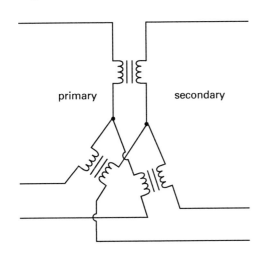

The turns ratio is most nearly

(A) 0.1
(B) 1
(C) 8
(D) 10

22. A transmission line has a per unit length inductance and capacitance of 2.5×10^{-6} H/m and 1.0×10^{-9} F/m, respectively, and is terminated with a purely resistive load. The standing wave ratio is measured at the load as 3.0. The load resistance is most nearly

(A) 25 Ω
(B) 50 Ω
(C) 100 Ω
(D) 150 Ω

23. What is the value of X at the completion of the following flow diagram?

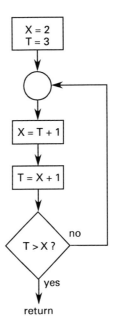

(A) 2
(B) 4
(C) 5
(D) The loop never ends.

24. A computer psuedocode program contains the following program segment.

$$X = 0$$
$$\text{FOR } T = -1 \text{ TO } 2$$
$$X = X + T$$
$$\text{NEXT } T$$

What is the value of X after the segment is executed?

(A) 2
(B) 3
(C) 4
(D) 6

25. An exponentially decaying voltage of the form $v(t)$ is applied to a digital circuit input. The trigger voltage is 2 V.

$$v(t) = 20e^{\frac{-t}{400\times 10^{-3}\,\text{s}}}$$

How much time is required for this voltage to decay to a value of 2 V?

(A) 200 ms
(B) 400 ms
(C) 920 ms
(D) 1800 ms

26. In the circuit shown, what is the current through the 6 Ω resistor?

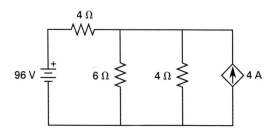

(A) 7.0 A
(B) 7.5 A
(C) 10 A
(D) 12 A

27. Consider the digital filter described by the z-plane pole-zero diagram shown.

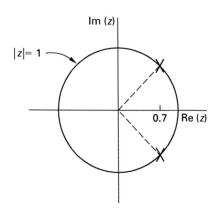

Which of the following statements is true?

(A) The filter is stable.
(B) The impulse response is periodic.
(C) The filter is non-causal.
(D) The filter is unstable.

28. Consider the circuit shown.

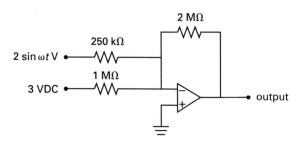

The voltage at the output is most nearly

(A) $\sin(\omega t - 180°)$
(B) $2\sin(\omega t - 180°) - 3$ V
(C) $16\sin\omega t + 6$ V
(D) $16\sin(\omega t - 180°) - 6$ V

29. A magnetic field of 0.0005 T makes a 30° angle as shown with a 1 m long wire carrying 0.5 A.

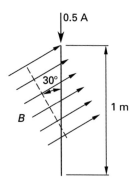

The force on the wire is most nearly

(A) 2.17×10^{-4} N into the paper
(B) 2.17×10^{-4} N out of the paper
(C) 5.00×10^{-4} N into the paper
(D) 5.00×10^{-4} N out of the paper

30. A common-emitter biasing circuit for a transistor amplifier operating in the active region is shown.

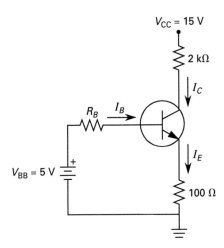

Assume the collector-base current ratio, β, equals 100, the temperature is 27°C, and the base-emitter voltage is 0.4 V. The value of R_B that results in a current gain of $g_m = 0.5\ \Omega^{-1}$ is most nearly

(A) 2 kΩ
(B) 5 kΩ
(C) 30 kΩ
(D) 40 kΩ

31. The figure shown represents a JFET amplifier. Assume the JFET is operating in the saturation region. The conductivity factor is 6.5×10^{-4} A/V².

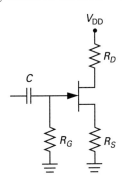

The current gain is $g_m = 2 \times 10^{-3}\ \Omega^{-1}$. The drain current I_D is most nearly

(A) 0.5 mA
(B) 1.0 mA
(C) 1.3 mA
(D) 1.5 mA

32. A transmission line has a characteristic impedance of 75 Ω and is connected to a load with a characteristic impedance of 377 Ω. The percentage of the signal voltage transmitted to the load is most nearly

(A) 10%
(B) 30%
(C) 50%
(D) 100%

33. Which logic function represents the minimal SOP of the following Karnaugh map? The X's represent "don't cares" and may be grouped with 0's or 1's.

CD \ AB	00	01	11	10
00	X	X		1
01		X	1	
11		1	1	
10	1			1

(A) $BD + \overline{B}\,\overline{D}$
(B) $B\overline{C}D + \overline{B}\,\overline{D}$
(C) $\overline{B}\,\overline{D}$
(D) $ABD + \overline{A}CD + \overline{B}\,\overline{D}$

Problems 34 and 35 are based on the following scenario. The active circuit shown has been on for a long time and is at steady state. The inductor and capacitor impedances are given for the frequency of the AC voltage source.

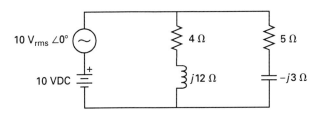

34. The rms voltage drop across the 4 Ω resistor is most nearly

(A) $10\ V_{rms} \angle -71.6°$
(B) $10\ VDC + 3.2\ V_{rms} \angle -71.6°$
(C) $3\ VDC + 3.2\ V_{rms} \angle -71.6°$
(D) $3.2\ V_{rms} \angle -18°$

35. The rms voltage drop across the 5 Ω resistor is most nearly

(A) $8.6\ V_{rms} \angle 31°$
(B) $10\ VDC + 8.6\ V_{rms} \angle 31°$
(C) $3\ VDC + 8.6\ V_{rms} \angle 31°$
(D) $6.2\ V_{rms} \angle 60°$

36. A control system has a characteristic equation of $s^3 + 2s^2 + 3s + K = 0$. For what values of K will the system be stable?

(A) $K < -6$
(B) $0 < K < 6$
(C) $0 < K$
(D) $6 < K$

Problems 37 and 38 are based on the circuit shown.

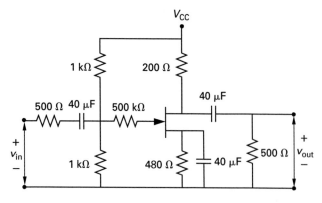

37. The input impedance at 100 kHz is most nearly

(A) 1 kΩ
(B) 2 kΩ
(C) 5 kΩ
(D) 50 kΩ

38. The purpose of the capacitors in the circuit is

(A) to speed up the voltage propagation
(B) to isolate the DC biasing circuit from the AC input and output
(C) to suppress noise form the DC voltage source V_{CC}
(D) to increase the amplification

39. An analog signal is sampled and processed digitally. The filtered signal is then sent through an output D/A converter where the analog signal is reconstructed.

What is the minimum sampling frequency required to accurately reconstruct the signal without aliasing?

(A) 60 Hz
(B) 120 Hz
(C) 190 Hz
(D) 380 Hz

40. Consider the following waveform.

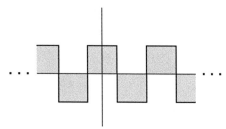

The ratio of the power in the Fourier series expansion element with $n = 3$ to the total power is most nearly

(A) $\dfrac{\pi}{\sqrt{2}}$
(B) $\dfrac{1}{\sqrt{2}}$
(C) $\dfrac{(3\pi)^2}{8}$
(D) $\dfrac{3\pi}{2}$

41. A system is governed by the following differential equations.
$$\dot{x}_2 + a_1 \dot{x}_1 + a_0 x_1 = u(t)$$
$$\dot{x}_1 = -x_2$$

The output of the system is
$$y_1(t) = 3x_1$$
$$y_2(t) = 4x_1 - 5\dot{x}_1$$

The output matrix of the state-variable control system model is

(A) $\begin{bmatrix} 3 & 0 \\ 4 & 5 \end{bmatrix}$

(B) $\begin{bmatrix} a_1 \\ a_0 \end{bmatrix}$

(C) $\begin{bmatrix} a_1 & 0 \\ 0 & a_0 \end{bmatrix}$

(D) $\begin{bmatrix} -5 & 3 \\ 4 & 0 \end{bmatrix}$

42. A capacitor is made with two equal conductive plates with free space between them. The capacitance is 1 pF. A dielectric material is placed between the capacitor plates. The new capacitance of the capacitor is 4 pF. The permittivity of the dielectric is most nearly

(A) 8.8×10^{-12} F/m
(B) 1.0×10^{-11} F/m
(C) 2.0×10^{-11} F/m
(D) 3.5×10^{-11} F/m

43. In the circuit shown, $X = 1$, $Y = 0$, and $Z = 1$.

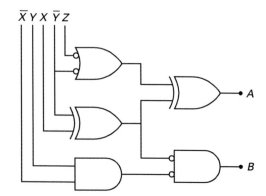

What are the outputs, A and B?

(A) $A = 0$, $B = 0$
(B) $A = 0$, $B = 1$
(C) $A = 1$, $B = 0$
(D) $A = 1$, $B = 1$

44. The concentration of holes in an n-type semiconductor is $p = 3.5 \times 10^5$ cm^{-3}, the intrinsic carrier concentration is $n_i = 1.5 \times 10^{10}$ cm^{-3}, and the electron mobility is $\mu_n = 1.5 \times 10^3$ cm^2/V·s. The resistivity of the semiconductor is most nearly

(A) 4.6×10^{-3} Ω·cm
(B) 5.0×10^{-3} Ω·cm
(C) 6.5 Ω·cm
(D) 10 Ω·cm

45. What is the POS representation equivalent to the SOP Boolean formula shown?

$$Y = A\overline{B}C + A\overline{B}\,\overline{C} + \overline{A}\,\overline{B}\,\overline{C} + \overline{A}\,\overline{B}C$$

(A) $Y = (A + \overline{B} + C)\cdot(\overline{A} + \overline{B})\cdot(\overline{B} + \overline{C})$
(B) $Y = (\overline{A} + B + \overline{C})\cdot(A + B)$
(C) $Y = (\overline{A} + B + \overline{C})\cdot(A + B)\cdot(\overline{A} + \overline{B} + C)$
(D) $Y = (\overline{A} + B + \overline{C})\cdot(A + \overline{B} + C)\cdot(B + C)$

46. A 40 MB hard disk is formatted to have 8 sectors and 1024 tracks. The fewest number of bytes needed to store a file on the disk is most nearly

(A) 1024 bytes
(B) 4882 bytes
(C) 5120 bytes
(D) 8192 bytes

47. The following table represents a 32-bit (single-precision) IEEE floating point number.

sign bit	8-bit signed exponent in Excess-127 representation	23-bit mantissa (the MSB is always 1 and therefore not shown, giving 24 bits of precision)
S	E'	M
0	0 1 1 1 1 1 1 0	0 1 1 0 0 0 ... 0 0 0 0

The value $F(B)$ represented by these bits is given by

$$F(B) = (\pm 1.M)(2)^{E' - 127}$$

What is the decimal equivalent of the floating point number shown?

(A) 1.100×10^{-1}
(B) 6.000×10^{-1}
(C) 6.875×10^{-1}
(D) 8.145×10^{-1}

48. A quarter-bridge circuit is used in an instrumentation system. The resistance of the sensor is proportional to the square of the phenomenon that is being measured by the sensor. The input voltage to the quarter-bridge circuit is constant. The output voltage of the quarter-bridge circuit is most nearly

(A) proportional to the phenomenon
(B) inversely proportional to the phenomenon
(C) proportional to the square of the phenomenon
(D) inversely proportional to the square of the phenomenon

49. A thermocouple measures which of the following?

(A) heat flux
(B) ratio of a temperature to a standard
(C) absolute temperature
(D) wind speed

50. A sinusoid signal is mixed with a carrier as shown. The modulated carrier is then filtered by the ideal filter shown.

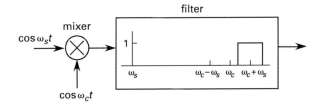

The output of the filter is most nearly

(A) $\frac{1}{2}(\cos\omega_s t + \cos\omega_c t)$
(B) $\frac{1}{2}(\cos\omega_s t - \omega_c t)$
(C) $\frac{1}{2}(\cos\omega_s t + \omega_c t)$
(D) $\frac{1}{2}((\cos\omega_s t + \omega_c t) - \cos(\omega_s t - \omega_c t))$

51. The waveform shown consists of an infinite series of identical pulses, each of 10 ms duration with a period of 1 s.

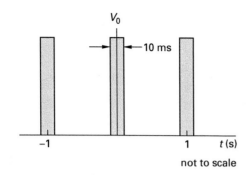

not to scale

The Fourier series expansion element with $n = 3$ has a magnitude of most nearly

(A) $0.001V_0$
(B) $0.01V_0$
(C) $0.02V_0$
(D) $0.03V_0$

52. At an equilibrium temperature of 50°C, the contact potential of a pn junction is 0.6 V. The intrinsic carrier concentration is $n_i = 1.5 \times 10^{10}$ cm^{-3}, and the donor concentration in the n-type material is $N_d = 7.0 \times 10^{17}$ cm^{-3}. The acceptor concentration is most nearly

(A) 7.0×10^{10} cm^{-3}
(B) 1.0×10^{11} cm^{-3}
(C) 7.4×10^{11} cm^{-3}
(D) 7.0×10^{17} cm^{-3}

53. The concept of "interleaving" is a method of arranging

(A) memory to increase size
(B) memory to increase speed
(C) data to decrease size
(D) data to increase speed

Problems 54 and 55 are based on the following control system diagram.

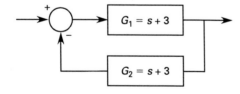

54. The time-domain transfer function for the control system shown is

(A) $e^{-3t} \sin t$
(B) $e^{-3t} \cos t$
(C) $\dfrac{t+3}{1+(t+3)^2}$
(D) $\dfrac{1}{1+(t+3)^2}$

55. The root locus diagram of the control system shown has

(A) a pole at $(-3, j0)$ and a zero at $(-3, j0)$
(B) a zero at $(-3, j0)$
(C) a double zero at $(-3, j0)$
(D) a double pole at $(-3, j0)$ and a zero at $(0, 0)$

56. What is most nearly the DC gain of the digital filter represented by the following difference equation?

$$y(k) = \tfrac{1}{5}(2x(k) + x(k-1) + 2x(k+1))$$

(A) 1
(B) 2
(C) 4
(D) 5

Problems 57 and 58 are based on the following circuit. $i_3 = 1 \text{ A} \angle 58°$, and $i_1 = 0.707 \text{ A} \angle -116°$.

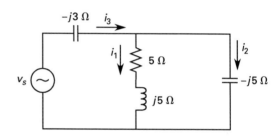

57. The impedance as seen by the voltage source is most nearly

(A) $4.3 \,\Omega \angle 8°$
(B) $8.0 \,\Omega \angle -50°$
(C) $9.4 \,\Omega \angle -58°$
(D) $11 \,\Omega \angle -68°$

58. The current i_2 is most nearly

(A) $1.7 \text{ A} \angle 60°$
(B) $1.9 \text{ A} \angle 160°$
(C) $2.0 \text{ A} \angle -60°$
(D) $2.0 \text{ A} \angle 60°$

Problems 59 and 60 are based on the following scenario.

An electromagnetic field of a uniform plane wave propagating in a dielectric nonmagnetic medium ($\mu = \mu_o$) is given by

$$E(t, z) = \mathbf{a}_x \sin\left(10^8 t - \dfrac{z}{\sqrt{3}}\right)$$

59. The relative permittivity is most nearly

(A) 1
(B) 2
(C) 3
(D) 4

60. The polarization of the wave is

(A) linear in the x direction
(B) linear in the y direction
(C) circular
(D) elliptical

SOLUTIONS

1. The circuit is a low-pass filter with a cutoff frequency of

$$\frac{1}{RC} = \frac{1}{(50\,\Omega)(200 \times 10^{-6}\,\text{F})}$$
$$= 100\,\text{rad/s}$$

The input is at the cutoff frequency. The cutoff frequency is the half-power point, so the amplitude of the output voltage is the input voltage divided by $\sqrt{2}$.

$$\frac{40\,\text{V}}{\sqrt{2}} = 28.3\,\text{V} \quad (28\,\text{V})$$

Alternately, the impedance of the capacitor is

$$X_C = -\frac{1}{\omega C} = -\frac{1}{\left(100\,\frac{\text{rad}}{\text{s}}\right)(200 \times 10^{-6}\,\text{F})}$$
$$= -50\,\Omega$$

The output amplitude is

$$v_{\text{in}}\left(\frac{|X_C|}{|R + jX_C|}\right) = v_{\text{in}}\left(\frac{|X_C|}{\sqrt{R^2 + X_C^2}}\right)$$
$$= (40\,\text{V})\left(\frac{50\,\Omega}{\sqrt{(50\,\Omega)^2 + (50\,\Omega)^2}}\right)$$
$$= 28.3\,\text{V} \quad (28\,\text{V})$$

The answer is C.

2. For convenience, reverse the number so the LSB is on the right.

$$1011001$$

To convert from two's complement to ordinary binary, take the complement and add 1.

$$1011001$$
$$0100110$$
$$\underline{+1}$$
$$0100111$$

The value of the word is

$$\left(\begin{array}{c}(1)(2)^5 + (0)(2)^4 + (0)(2)^3 \\ +(1)(2)^2 + (1)(2)^1 + (1)(2)^0\end{array}\right)(2.4\,\text{m})$$
$$= 93.6\,\text{m} \quad (94\,\text{m})$$

The answer is D.

3. The neutral current must satisfy Kirchoff's current law.

$$I_N \equiv I_A + I_B + I_C$$
$$= \frac{V_A}{Z_A} + \frac{V_B}{Z_B} + \frac{V_C}{Z_C}$$
$$= \frac{480\angle 0°}{50\,\Omega} + \frac{480\,\text{V}\,\angle 120°}{40 + j30\,\Omega} + \frac{480\,\text{V}\,\angle 240°}{43.3 + j25\,\Omega}$$
$$= 9.6\angle 0° + \frac{480\,\text{V}\angle 120°}{50\,\Omega\angle 36.87°} + \frac{480\,\text{V}\angle 240°}{50\,\Omega\angle 30°}$$
$$= 9.6\angle 0° + 9.6\angle 83.13° + 9.6\angle 210°$$
$$= 5.32\,\text{A}\angle 62.77° \quad (5.3\,\text{A}\angle 62.8°)$$

The answer is B.

4. The maximum voltage across the capacitor is 22 V, so using a 50% derating means the capacitor must have a rating of at least 44 V. Answer (C) will not work. The largest voltage across the resistor is $29\,\text{V} - 18\,\text{V} = 11\,\text{V}$, so the power dissipated is

$$\frac{V^2}{R} = \frac{(11\,\text{V})^2}{500\,\Omega} = 0.24\,\text{W}$$

To meet the 50% derating requirements, the resistor must have a value of at least 0.48 W. Answer (B) increases the power dissipated by the resistor. The wattage rating of answer (A) meets the derating requirement.

The answer is A.

5. Use the binary number system code table from the NCEES Handbook to convert the binary and the hexadecimal numbers in the numerator to octal. That gives

$$1101_2 = 15_8$$
$$A_{16} = 12_8$$

Calculate the value of the octal numbers and convert this number to decimal.

$$15_8 - 12_8 + 34_8 = 37_8$$
$$37_8 = (3)(8)^1 + (7)(8)^0 = 31_{10}$$

Convert the denominator to decimal and calculate the fraction.

$$22_3 = (2)(3)^1 + (2)(3)^0$$
$$= 8_{10}$$
$$\frac{31_{10} + 17_{10}}{8_{10}} = 6_{10}$$

From the binary number system code table,

$$6_{10} = 0110_{BCD}$$

The answer is A.

6. The gain and phase margins are shown graphically as follows.

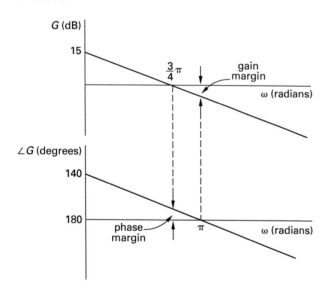

The gain margin is given by

$$GM = -20 \log_{10}(|G(j\omega_{180°})|)$$

The slope of the line is

$$\frac{15}{\frac{3}{4}\pi} = \frac{20}{\pi} \text{ dB/rad}$$

So,

$$G(j\omega_{180°}) = -5 \text{ dB}$$

This gain is shown in decibels, so no conversion is necessary. As shown in the gain margin equation, the gain margin is the negative of the gain at $\omega = 180°$, so

$$GM = 5 \text{ dB}$$

The phase margin is given by

$$PM = 180° + \angle G(j\omega_{0 \text{ dB}})$$

The slope of the line is $40/\pi$ degrees/rad. So,

$$\angle G(j\omega_{0 \text{ dB}}) = -170°$$
$$PM = 180° + \angle G(j\omega_{0 \text{ dB}})$$
$$= 180° + -170°$$
$$= 10°$$

The answer is B.

7. The synchronous speed is given by

$$n_s = (120)\left(\frac{f}{p}\right) \quad \text{[in rpm]}$$

There are three pole pairs, so the number of poles is $p = 6$.

$$n_s = (120)\left(\frac{f}{p}\right) = (120)\left(\frac{60 \text{ rpm}}{6}\right)$$
$$= 1200 \text{ rpm}$$

The slip is given by

$$s = \frac{n_s - n}{n_s}$$
$$= \frac{1200 \text{ rpm} - 1080 \text{ rpm}}{1200 \text{ rpm}}$$
$$= 0.1$$

The slip is a unitless ratio of the synchronous speed so the angular velocity of the slip is the slip times the synchronous speed in rad/s.

$$(0.1)\left(1200 \, \frac{\text{rotation}}{\text{min}}\right)\left(2\pi \, \frac{\text{rad}}{\text{rotation}}\right)\left(\frac{1 \text{ min}}{60 \text{ s}}\right)$$
$$= 4\pi \text{ rad/s}$$

The answer is D.

8. The problem can be simplified by converting the current source and its 3 Ω resistor to the Thevenin equivalent.

$$V_{oc} = R_{Th} I_{\text{short circuit}} = (3 \text{ Ω})(4 \text{ A})$$
$$= 12 \text{ V}$$

The circuit simplfies to become

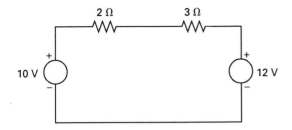

The current through the 2 Ω resistor is

$$i_{2\,\Omega} = \frac{12\text{ V} - 10\text{ V}}{2\text{ }\Omega + 3\text{ }\Omega} = 0.4\text{ A}$$

Because the 12 V source is greater than the 10 V source, this current is from right to left.

The answer is A.

9. The time-invariant convolution is given by

$$x(t) * y(t) = \int_{-\infty}^{\infty} x(\tau) y(t-\tau) d\tau$$
$$= \int_{-\infty}^{\infty} x(t) e^{5(t-\tau)} d\tau$$
$$= e^{5t} \int_{-\infty}^{\infty} x(t) e^{-5\tau} d\tau$$
$$= e^{5t} \int_{0}^{1} 5 e^{-5\tau} d\tau = e^{5t} \left. -\frac{5 e^{-5\tau}}{5} \right|_{0}^{1}$$
$$= e^{5t} \left(-e^{-5} + 1 \right)$$
$$= e^{5t} \left(1 - e^{-5} \right)$$

The answer is C.

10. The input is $S = 1$ because $Q_1 = 0$, so the next state of $Q_1 = 1$.

The present state is $Q_1 = 0$ and $Q_2 = 0$, so the exclusive OR condition is not met, and $D = 0$. The next state of $Q_2 = 0$.

The answer is C.

11. The Karnaugh map for segment g is

CD\AB	00	01	11	10
00	0	1	X	1
01	0	1	X	1
11	1	0	X	X
10	1	1	X	X

The two-square group in the upper left has the maxterm Boolean formula $A + B + C$ because both conditions of D are present and A, B, and C are false. The two-square group in the middle has the maxterm Boolean formula $\overline{B} + \overline{C} + \overline{D}$ because both conditions of A are present and B, C, and D are true. The POS Boolean formula is

$$(A + B + C) \cdot (\overline{B} + \overline{C} + \overline{D})$$

The answer is D.

12. The phase of the signal is given by

$$\phi_i(t) = \omega_c t + k_f \int_{-\infty}^{t} m(\tau) d\tau$$
$$m(\tau) = \sin \omega_m \tau$$
$$\int_{-\infty}^{t} (\sin \omega_m \tau) d\tau = -\frac{\cos \omega_m t}{\omega_m}$$
$$\phi_i(t) = \omega_c t - \frac{k_f \cos \omega_m t}{\omega_m}$$

The answer is A.

13. The instantaneous frequency is given by

$$\omega_i = \omega_c + m(t) = \omega_c t + k_f t \sin \omega_m t$$

The answer is B.

14. After the filter, the modulation Laplace transform looks like answer (A). Superimposing the spectrum onto the carrier duplicates this spectrum around both plus the carrier frequency and minus the carrier frequency. This is the spectrum seen in answer (C). Note that answer (D) must be wrong because power spectral density must always be positive.

The answer is C.

15. The waveform total excursion is 10 V. The 10% and 90% points are 1 V and 9 V, respectively. The time between these two points is the rise time. First, compute the time required to reach 1 V.

$$1\text{ V} = 10\text{ V} - (10\text{ V}) e^{\frac{-t}{200\text{ ns}}}$$
$$-9\text{ V} = -(10\text{ V}) e^{\frac{-t}{200\text{ ns}}}$$
$$0.9 = e^{\frac{-t}{200\text{ ns}}}$$
$$\ln 0.9 = \ln e^{\frac{-t}{200\text{ ns}}}$$
$$-0.10536 = \left(\frac{-t}{200\text{ ns}} \right) \ln e$$
$$t_{1\text{ V}} = 21.07\text{ ns}$$

Repeat for the 90% value, 9 V.

$$9 \text{ V} = 10 \text{ V} - (10 \text{ V})e^{\frac{-t}{200 \text{ ns}}}$$

$$-1 \text{ V} = -(10 \text{ V})e^{\frac{-t}{200 \text{ ns}}}$$

$$0.1 = e^{\frac{-t}{200 \text{ ns}}}$$

$$\ln 0.1 = \ln e^{\frac{-t}{200 \text{ ns}}}$$

$$-2.3026 = \left(\frac{-t}{200 \text{ ns}}\right) \ln e$$

$$t_{9 \text{ V}} = 460.5 \text{ ns}$$

$$\text{rise time} = t_{9 \text{ V}} - t_{1 \text{ V}} = 460.5 \text{ ns} - 21.07 \text{ ns}$$
$$= 439.4 \text{ ns} \quad (440 \text{ ns})$$

The answer is D.

16. The voltage gain of an operational amplifier circuit in this configuration is the negative of the ratio of the magnitude of the feedback impedance over the input impedance.

$$G = -\frac{|Z_{\text{feedback}}|}{|Z_{\text{in}}|}$$

$$Z_{\text{in}} = jX_C = -\frac{j}{2\pi fC}$$
$$= -\frac{j}{(2\pi)(100 \text{ Hz})(0.001 \times 10^{-6} \text{ F})}$$
$$= -j\,1.59 \times 10^6 \text{ }\Omega$$

$$Z_{\text{feedback}} = \frac{-jX_C R}{R - jX_C}$$

Since X_C is the same for the input and feedback, it can be substituted in as follows.

$$Z_{\text{feedback}} = \frac{(-j\,1.59 \times 10^6 \text{ }\Omega)(10 \times 10^6)}{10 \times 10^6 - j\,1.59 \times 10^6 \text{ }\Omega}$$
$$= \left(\frac{-j\,1.59 \times 10^{13} \text{ }\Omega}{10 \times 10^6 - j\,1.59 \times 10^6 \text{ }\Omega}\right)$$
$$\times \left(\frac{10 \times 10^6 + j\,1.59 \times 10^6 \text{ }\Omega}{10 \times 10^6 + j\,1.59 \times 10^6 \text{ }\Omega}\right)$$
$$= \frac{-2.53 \times 10^{19} - j\,1.59 \times 10^{20} \text{ }\Omega}{10^{14} + 2.53 \times 10^{12}}$$
$$= -2.47 \times 10^5 - j1.55 \times 10^6 \text{ }\Omega$$

The minus signs affect only the phase, so they can be dropped.

$$\frac{|Z_{\text{feedback}}|}{|Z_{\text{in}}|} = \frac{\sqrt{(2.47 \times 10^5 \text{ }\Omega)^2 + (1.55 \times 10^6 \text{ }\Omega)^2}}{1.59 \times 10^6 \text{ }\Omega}$$
$$= 0.99 \quad (1.0)$$

The answer is A.

17. Maximum power transfer occurs when the impedance seen at the primary side of the transformer matches the line impedance, Z_1. Calculating the secondary impedance and then reflecting it to the primary side results in

$$Z_s = Z_2 \| (Z_3 + Z_4) = \frac{Z_2(Z_3 + Z_4)}{Z_2 + (Z_3 + Z_4)}$$

$$Z_p = a^2 Z_S = \left(\frac{1}{N}\right)^2 \left(\frac{Z_2(Z_3 + Z_4)}{Z_2 + Z_3 + Z_4}\right)$$

Setting $Z_p = Z_1$ yields

$$\left(\frac{1}{N}\right)^2 \left(\frac{Z_2(Z_3 + Z_4)}{Z_2 + Z_3 + Z_4}\right) = Z_1$$

$$\left(\frac{Z_2}{Z_1}\right)\left(\frac{Z_3 + Z_4}{Z_2 + Z_3 + Z_4}\right) = N^2$$

$$N = \sqrt{\frac{Z_2(Z_3 + Z_4)}{Z_1(Z_2 + Z_3 + Z_4)}}$$

The answer is C.

18. The circuit is a half-wave rectifier.

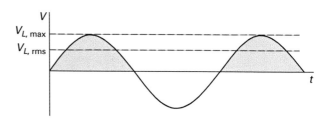

The power produced at the load, P_L, is proportional to the rms value of the voltage.

$$V_{L,\text{rms}} = \frac{V_{L,\text{max}}}{2}$$

$$P_L = (I_{L,\text{rms}})(V_{L,\text{rms}})$$
$$= \left(\frac{V_{L,\text{max}}}{200 \text{ }\Omega}\right)(V_{L,\text{rms}})$$
$$= \frac{V_{L,\text{rms}}^2}{200 \text{ }\Omega}$$
$$= \frac{\left(\frac{V_{L,\text{max}}}{2}\right)^2}{200 \text{ }\Omega}$$
$$= \frac{V_{L,\text{max}}^2}{800 \text{ }\Omega}$$
$$= 25 \text{ W}$$

$$V_{L,\text{max}} = \sqrt{(25 \text{ W})(800 \text{ }\Omega)}$$
$$= 141.4 \text{ V} \quad (\text{at the load})$$

Because the shunt resistance creates a voltage divider,

$$V_{L,\max} = V_m \left(\frac{200 \, \Omega}{20 \, \Omega + 200 \, \Omega} \right)$$
$$= 0.909 \, V_m$$
$$V_m = \frac{V_{L,\max}}{0.909}$$
$$= \frac{141.4 \, \text{V}}{0.909}$$
$$= 155.6 \, \text{V} \quad (160 \, \text{V})$$

The answer is D.

19. When the formula A1 + A1 is copied into cell C1, it becomes A1 + B1. The number displayed in cell B1 is $2 + 2 = 4$. The number displayed in cell C1 is $2 + 4 = 6$.

The answer is C.

20. The exclusive OR is an associative operator, as are all the Boolean operators, so

$$W \oplus X \oplus Y \oplus Z = (W \oplus X) \oplus (Y \oplus Z)$$

The answer is A.

21. This is a wye configuration, so the phase voltage for the secondary is obtained by

$$V_L = \sqrt{3} V_{\text{phase}}$$
$$V_{\text{phase}} = \frac{V_L}{\sqrt{3}} = \frac{208 \, \text{V}}{\sqrt{3}}$$
$$= 120 \, \text{V}$$

The turns ratio is the magnitude of the primary phase voltage divided by the magnitude of the secondary phase voltage.

$$\frac{|V_p|}{|V_s|} = \frac{1.2 \times 10^3 \, \text{V}}{120 \, \text{V}}$$
$$= 10$$

The answer is D.

22. By definition, the standing wave ratio, SWR, is given by $(1 + |\Gamma|)/(1 - |\Gamma|)$, where Γ is the reflection coefficient. Since the SWR is given, it is necessary to find Γ in terms of SWR.

$$\text{SWR}\,(1 - |\Gamma|) = 1 + |\Gamma|$$
$$\text{SWR} - \text{SWR}\,|\Gamma| = 1 + |\Gamma|$$
$$\text{SWR} - 1 = \text{SWR}\,|\Gamma| + |\Gamma|$$
$$= (\text{SWR} + 1)\,|\Gamma|$$
$$\Gamma = \frac{\text{SWR} - 1}{\text{SWR} + 1} = \frac{3.0 - 1}{3.0 + 1}$$
$$= 0.5$$

Likewise, the reflection coefficient is given by

$$\Gamma = \frac{Z_L - Z_0}{Z_L + Z_0}$$
$$\Gamma (Z_L + Z_0) = Z_L - Z_0$$
$$Z_0 (1 + \Gamma) = Z_L (1 - \Gamma)$$

Therefore

$$Z_L = Z_0 \left(\frac{1 + \Gamma}{1 - \Gamma} \right) = \sqrt{\frac{L}{C}} \left(\frac{1 + \Gamma}{1 - \Gamma} \right)$$
$$= \sqrt{\frac{2.5 \times 10^{-6} \, \frac{\text{H}}{\text{m}}}{1.0 \times 10^{-9} \, \frac{\text{F}}{\text{m}}}} \left(\frac{1 + 0.5}{1 - 0.5} \right)$$
$$= (50 \, \Omega) \left(\frac{1.5}{0.5} \right)$$
$$= 150 \, \Omega$$

The answer is D.

23. During the first pass through the loop, the program performs the following operations:

$$X = T + 1 = 3 + 1 = 4$$

The next operation is

$$T = X + 1 = 4 + 1 = 5$$

Since the condition T > X is satisfied, the program segment is over and X = 4.

The answer is B.

24. The FOR statement requires the loop be repeated four times, when $T = -1$, $T = 0$, $T = 1$, and $T = 2$.

The sequence of operations is

$$X = X + T = 0 - 1 = -1$$
$$X = X + T = -1 + 0 = -1$$
$$X = X + T = -1 + 1 = 0$$
$$X = X + T = 0 + 2 = 2$$

The answer is A.

25. $$v(t) = (20 \text{ V})e^{\frac{-t}{400 \times 10^{-3} \text{ s}}}$$

Substitute and divide both sides of the equation by 20.

$$\frac{2 \text{ V}}{20} = \frac{(20 \text{ V})e^{\frac{-t}{400 \times 10^{-3} \text{ s}}}}{20}$$

$$0.1 = e^{\frac{-t}{400 \times 10^{-3} \text{ s}}}$$

Take the natural logarithm of both sides of the equation.

$$\ln 0.1 = \ln e^{\frac{-t}{400 \times 10^{-3} \text{ s}}} = -\frac{t}{400 \times 10^{-3} \text{ s}}$$

Solve for t.

$$t = (-400 \times 10^{-3} \text{ s}) \ln 0.1$$
$$= (-400 \times 10^{-3} \text{ s})(-2.303)$$
$$= 0.921 \text{ s} \quad (920 \text{ ms})$$

The answer is C.

26. The current through the 6 Ω resistor can be found through superposition of the current due to the voltage source and the current due to the current source. First, replace the current source with an open circuit, as shown.

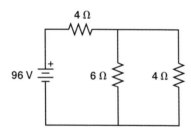

The total resistance is

$$4 \text{ }\Omega + \frac{(4 \text{ }\Omega)(6 \text{ }\Omega)}{4 \text{ }\Omega + 6 \text{ }\Omega} = 6.4 \text{ }\Omega$$

The total current is

$$\frac{96 \text{ V}}{6.4 \text{ }\Omega} = 15 \text{ A}$$

The current through the 6 Ω resistor due to the voltage source is

$$\left(\frac{4 \text{ }\Omega}{4 \text{ }\Omega + 6 \text{ }\Omega}\right)(15 \text{ A}) = 6 \text{ A}$$

Next, replace the voltage source with a short circuit in the original circuit.

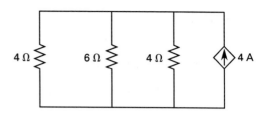

The two 4 Ω resistors in parallel are equivalent to a 2 Ω resistor, so the current through the 6 Ω resistor due to the current source is

$$\left(\frac{2 \text{ }\Omega}{6 \text{ }\Omega + 2 \text{ }\Omega}\right)(4 \text{ A}) = 1 \text{ A}$$

The total current through the 6 Ω resistor is

$$6 \text{ A} + 1 \text{ A} = 7 \text{ A} \quad (7.0 \text{ A})$$

The answer is A.

27. Since the poles are located on the unit circle, the filter is considered neither stable nor unstable, rather *marginally* stable.

Since there are more poles (two) than zeros (none), the filter is causal.

However, since the poles are imaginary (i.e., they make an angle with the real axis), the impulse response is oscillatory, and in this case, since they are on the unit circle ($|z| = 1$), the response neither decays nor grows. Therefore, it is periodic.

The answer is B.

28. The current through the feedback resistor must be equal and opposite to the two input currents to keep the voltage at the inverting input (−) of the operational amplifier equal to zero. The current input to the (−) input is

$$\frac{2 \sin \omega t \text{ V}}{250 \times 10^3 \text{ }\Omega} + \frac{3 \text{ V}}{10^6 \text{ }\Omega} = (8 \times 10^{-6} \text{ A}) \sin \omega t + (3 \times 10^{-6} \text{ A})$$

The feedback voltage must be

$$((-8 \times 10^{-6} \text{ A}) \sin \omega t + (3 \times 10^{-6} \text{ A}))(2 \times 10^6 \text{ }\Omega)$$
$$= -16 \sin \omega t - 6 \text{ V}$$
$$= 16 \sin(\omega t - 180°) - 6 \text{ V}$$

Note that the minus sign on the sine function is equivalent to a 180° phase change.

The answer is D.

29. The angle between the line perpendicular to the magnetic flux density and the wire is 30°, so the angle between the vector of the current and the magnetic flux density is $30° + 90° = 120°$. The force on a current carrying conductor is

$$|\mathbf{F}| = I|\mathbf{L} \times \mathbf{B}| = I|\mathbf{L}||\mathbf{B}|\sin 120°$$
$$= (0.866 \text{ A})(1 \text{ m})(0.0005 \text{ T})(0.5)$$
$$= 2.17 \times 10^{-4} \text{ N}$$

Using the right hand with the page horizontal, point your fingers in the direction of the current and curl your fingers in the direction of the magnetic field. Your thumb is pointing up, so the direction is out of the paper.

The answer is B.

30. Calculate the approximate gain of the transistor.

$$g_m \approx \frac{I_C}{V_T}$$

$$T = 27°C + 273.15° = 300.15 \text{K}$$

The thermal voltage is given by

$$V_T = \frac{kT}{Q} = \frac{\left(1.38 \times 10^{-23} \frac{\text{J}}{\text{K}}\right)(300.15\text{K})}{1.602 \times 10^{-19} \text{ C}}$$
$$= 0.0259 \text{ V}$$

The desired collector current is

$$I_C \approx g_m V_T = (0.5)(0.0259 \text{ A}) = 0.013 \text{ A}$$

The desired base current is

$$I_B = \frac{I_C}{\beta} = \frac{0.013 \text{ A}}{100} = 1.3 \times 10^{-4} \text{ A}$$

Taking the KVL around the base-emitter loop,

$$V_{BB} = R_B I_B + V_{BE} + R_E I_E$$
$$= R_B I_B + V_{BE} + R_E(I_B + I_C)$$

$$R_B = \frac{V_{BB} - V_{BE} - R_E(I_B + I_C)}{I_B}$$

$$= \frac{5 \text{ V} - 0.4 \text{ V} - (100\,\Omega)}{1.3 \times 10^{-4} \text{ A}}$$
$$= 25.3 \text{ k}\Omega \quad (30 \text{ k}\Omega)$$

The answer is C.

31. The current gain is given by

$$g_m = \frac{2\sqrt{I_{\text{DSS}} I_D}}{|V_p|}$$

Solving for I_D,

$$I_D = \frac{g_m^2 V_p^2}{4 I_{\text{DSS}}}$$

Use the definition of the conductivity factor, $I_{\text{DSS}} = KV_p^2$.

$$I_D = \frac{g_m^2 V_p^2}{4KV_p^2}$$

$$= \frac{(2 \times 10^{-3} \, \Omega^{-1})^2}{(4)\left(6.5 \times 10^{-4} \, \dfrac{\text{A}}{\text{V}^2}\right)}$$

$$= 1.54 \times 10^{-3} \text{A} \quad (1.5 \text{ mA})$$

The answer is D.

32. The fraction of the voltage transmitted is one minus the reflection coefficent, so the percentage transmitted $= (1 - \rho) \times 100\%$.

$$\rho = \left|\frac{Z_L - Z_o}{Z_L + Z_o}\right|$$
$$= \frac{377\,\Omega - 75\,\Omega}{377\,\Omega + 75\,\Omega}$$
$$= 0.668$$

$$\% \text{ transmitted} = (1 - 0.668) \times 100\%$$
$$= 33.2\% \quad (30\%)$$

The answer is B.

33. The Karnaugh map minimum SOP grouping is

CD \ AB	00	01	11	10
00	X	X		1
01		X	1	
11		1	1	
10	1			1

The grouping of the four center squares uses one of the "don't cares" to make a group of four. It has the minterm Boolean formula BD because both conditions of A and C are present and B and D are true. One of the other "don't cares" is used to make a group of four with the four corners. It has the minterm Boolean formula $\overline{B}\,\overline{D}$ because both conditions of A and C are present and B and D are false. The minimum Boolean POS representation is

$$BD + \overline{B}\,\overline{D}$$

The answer is A.

34. The voltage on the resistor is a superposition of the DC and AC voltages. The inductor is effectively a short circuit to the DC voltage, so the DC voltage on the 4 Ω resistor is 10 VDC. The voltage divider concept can be used with impedances instead of resistances to determine the voltage across the resistor.

$$V_R = \left(\frac{Z_R}{Z_t}\right)V$$
$$= \left(\frac{4}{4+j12}\right)(10\ V_{rms}\angle 0°)$$
$$= \left(\frac{40}{4+j12}\right)\left(\frac{4-j12}{4-j12}\right) V_{rms}\angle 0°$$
$$= \frac{(40)(4-j12)}{160} V_{rms}$$
$$= 1-j3\ V_{rms}$$
$$= 3.2\ V_{rms}\angle -71.6°$$

The total rms voltage on the resistor is

$$10\ V_{DC} + 3.2\ V_{rms}\angle -71.6°$$

The answer is B.

35. The capacitor is effectively an open circuit to the DC voltage, so the voltage on the resistor is the AC voltage divider in impedance.

$$V_R = \left(\frac{Z_R}{Z_t}\right)V$$
$$= \left(\frac{5}{5-j3}\right)(10\ V_{rms}\angle 0°)$$
$$= \left(\frac{50}{5-j3}\right)\left(\frac{5+j3}{5+j3}\right) V_{rms}\angle 0°$$
$$= \frac{(50)(5+j3)}{34} V_{rms}$$
$$= 7.35 + 4.41\ V_{rms}$$
$$= 8.57\ V_{rms}\angle 31°\quad (8.6\ V_{rms}\angle 31°)$$

The answer is A.

36. Use the Routh criteria to develop the following table.

$a_3 = 1$	$a_1 = 3$
$a_2 = 2$	$a_0 = K$
$b_1 = \frac{(2)(3) - 1K}{2}$ $= 3 - \frac{K}{2}$	$b_2 = \frac{(2)(0) - (1)(0)}{2}$ $= 0$
$c_1 = \frac{K\left(3 - \frac{K}{2}\right) - (2)(0)}{3 - \frac{K}{2}}$ $= K$	0
0	0

Therefore, the control system is only stable if both of the following conditions are true.

$$K > 0$$
$$3 - \frac{K}{2} > 0$$

The second condition is equivalent to

$$K < 6$$

Both conditions together is equivalent to

$$0 < K < 6$$

The answer is B.

37. The impedance of the capacitors is

$$X_C = -\frac{1}{\omega C}$$
$$= -\frac{1}{(2\pi)\left(100 \times 10^3\ \frac{rad}{s}\right)(40 \times 10^{-6}\ F)}$$
$$= -0.04\ \Omega$$

So, the reactance of the capacitors can be ignored.

The current into the gate is insignificant, so the 500 kΩ resistor can be ignored. The input impedance is, therefore, the 500 Ω resistor in series with the two 1 kΩ resistors in parallel.

$$500\ \Omega + \frac{(1000\ \Omega)(1000\ \Omega)}{1000\ \Omega + 1000\ \Omega} = 1000\ \Omega \quad (1\ k\Omega)$$

The answer is A.

38. The capacitors allow the DC bias on the transistor to operate the transistor in the desired range without the DC voltage affecting the input or the output. Therefore, the capacitors isolate the DC biasing circuit from the AC input and output.

The answer is B.

39. The signal $x(t)$ generated by the oscillator can be determined by inspection of the Laplace transform table.

$$x(t) = sin 377t = \sin 2\pi f t = \sin 2\pi 60 t$$

The minimum sampling frequency, f_s, (Nyquist criterion) required to reconstruct $x(t)$ from $x(n)$ is

$$f_s \geq 2f = (2)(60\ Hz) = 120\ Hz$$

The answer is B.

40. The power in the third harmonic ($n = 3$) is proportional to the rms value ($P = V_{rms}^2 R$).

Since the harmonic is sinusoidal, the rms value is

$$V_{rms} = \frac{V_{max,harmonic}}{\sqrt{2}}$$

The maximum value of the third harmonic is

$$V_{max,harmonic} = \frac{4V_o}{n\pi} = \frac{4V_o}{3\pi}$$

The rms value of the original signal is equal to the maximum value, so the power is proportional to the square of the maximum voltage. The ratio of the powers is

$$\left(\frac{V_o^2}{V_o^2}\right)\left(\frac{\sqrt{2}(3\pi)}{4}\right)^2 = \frac{(3\pi)^2}{8}$$

The answer is C.

41. The vector notation for this system is

$$y_1(t) = 3x_1$$
$$y_2(t) = 4x_1 - 5\dot{x}_1 = 4x_1 + 5x_2$$
$$\begin{bmatrix} \dot{x}_1 \\ \dot{x}_2 \end{bmatrix} = \begin{bmatrix} 0 & 1 \\ -a_0 & -a_1 \end{bmatrix}\begin{bmatrix} x_1 \\ x_2 \end{bmatrix} + \begin{bmatrix} 0 \\ 1 \end{bmatrix}u(t)$$
$$\begin{bmatrix} y_1 \\ y_2 \end{bmatrix} = \begin{bmatrix} 3 & 0 \\ 4 & 5 \end{bmatrix}\begin{bmatrix} x_1 \\ x_2 \end{bmatrix} + \begin{bmatrix} 0 \\ 0 \end{bmatrix}u(t)$$
$$\mathbf{A} = \begin{bmatrix} 0 & 1 \\ -a_0 & -a_1 \end{bmatrix}$$
$$\mathbf{B} = \begin{bmatrix} 0 \\ 1 \end{bmatrix}$$
$$\mathbf{C} = \begin{bmatrix} 3 & 0 \\ 4 & 5 \end{bmatrix}$$
$$\mathbf{D} = \begin{bmatrix} 0 \\ 0 \end{bmatrix}$$

The output matrix is **C**.

The answer is A.

42. The capacitance of a capacitor is calculated as

$$C = \frac{\varepsilon A}{r}$$

The area (A) and the distance between the plates (r) do not change, so only the permittivity changes.

$$C_{new} = 4C_{old} = \frac{\varepsilon A}{r}$$
$$= (4)\left(\frac{\varepsilon_o A}{r}\right)$$

$$\varepsilon = 4\varepsilon_o = (4)\left(8.854 \times 10^{-12}\,\frac{F}{m}\right)$$
$$= 3.54 \times 10^{-11}\,F/m \quad (3.5 \times 10^{-11}\,F/m)$$

The answer is D.

43. Since the NAND gate at the top left has inputs 1 and 1, the condition is not satisfied; therefore, the output is 0. Since the XOR gate at the middle left has inputs 1 and 1, the condition is not satisfied; therefore, the output is 0. Since the AND gate at the bottom left has inputs 0 and 0, the condition is not satisfied; therefore, the output is 0. Since the XOR gate at the top right has inputs 0 and 0, the condition is not satisfied; therefore, the output is $A = 0$. Since the NOR gate at the bottom right has inputs 0 and 0, the condition is satisfied; therefore, the output is $B = 1$.

The answer is B.

44. The holes are the minority carrier in an n-type semiconductor and contribute insignificantly to the conductivity. The free electron concentration can be estimated from the hole concentration.

$$pn = n_i^2$$
$$n = \frac{n_i^2}{p} = \frac{\left(1.5 \times 10^{10}\,\frac{1}{cm^3}\right)^2}{3.5 \times 10^5\,\frac{1}{cm^3}}$$
$$= 6.43 \times 10^{14}\,\frac{1}{cm^3} \approx N_d$$

The conductivity can be derived from

$$\sigma \approx q(N_d\mu_n)$$
$$\rho = \frac{1}{\sigma} \approx \frac{1}{qN_d\mu_n}$$
$$= \frac{1}{(1.602 \times 10^{-19}\,C)\left(6.43 \times 10^{14}\,\frac{1}{cm^3}\right)}$$
$$\times \left(1.5 \times 10^3\,\frac{cm^2}{V\cdot s}\right)$$
$$= 6.47\,\Omega\cdot cm \quad (6.5\,\Omega\cdot cm)$$

The answer is C.

45. The Boolean formula contains the minterms 101, 100, 000, and 001.

$$Y = A\cdot\overline{B}\cdot C + A\cdot\overline{B}\cdot\overline{C} + \overline{A}\cdot\overline{B}\cdot\overline{C} + \overline{A}\cdot\overline{B}\cdot C$$

The POS form must contain the maxterms 010, 011, 111, and 110. (Note that there are a total of 2^n terms, where n is the number of variables.) Both conditions of A are present in the maxterms 011 and 111, so they can be combined in the term $(A+\overline{B}+\overline{C}) \cdot (\overline{A}+\overline{B}+\overline{C}) = \overline{B}+\overline{C}$. Similarly, $(\overline{A}+\overline{B}+C) \cdot (\overline{A}+\overline{B}+\overline{C}) = \overline{A}+\overline{B}$. The combined result is

$$Y = (A+\overline{B}+C) \cdot (\overline{A}+\overline{B}) \cdot (\overline{B}+\overline{C})$$

The answer is A.

46. Hard disks are accessed by sectors and tracks. This one has $N = (8)(1024) = 8192$ locations.

The total capacity is 40 MB, which equals $(40)(2)^{20} = 41{,}943{,}040$ bytes. Compute the smallest segment needed to store a file on the disk.

$$= \frac{41{,}943{,}040 \text{ bytes}}{8192 \text{ locations}} = 5120 \text{ bytes/location}$$

The answer is C.

47. Use the bitmap provided.

$$S = 0 \ (+)$$
$$E' = 0\,1\,1\,1\,1\,1\,1\,0$$
$$= 126_{10}$$
$$M = 0\,1\,1\,0\,0\,0\,0\,0...0\,0\,0$$
$$F(B) = (\pm 1.M)(2)^{E'-127}$$
$$= (1.011000...000)(2)^{126-127}$$
$$= (1.01100)(2)^{-1}$$
$$= 0.101100$$
$$= 0.6875_{10}$$
$$= 6.875 \times 10^{-1}$$

The answer is C.

48. The equation for the output of a quarter bridge circuit is given by

$$V_{\text{out}} = V_{\text{in}} \left(\frac{\Delta R}{4R}\right)$$

The output voltage is proportional to the change in resistance, which is proportional to the square of the phenomenon.

$$\Delta R \propto (\text{phenomenon})^2$$

Substituting,

$$E_{\text{out}} \propto E_{\text{in}}\left(\frac{(\text{phenomenon})^2}{4R}\right)$$

Since R and E_{in} are constants,

$$E_{\text{out}} \propto (\text{phenomenon})^2$$

The answer is C.

49. Thermocouples are two wires of dissimilar metals joined only at the point being measured and at a point with a standard temperature (usually a known temperature). The current is related to the ratio of the temperature at the point being measured and the standard temperature. Thermocouples are not thermometers; their outputs represent a ratio to the standard, not the absolute temperature.

The answer is B.

50. The output of the mixer is

$$\cos \omega_s t \cos \omega_c t$$

Using a trigonometric identity,

$$\cos \omega_s t \cos \omega_c t = \tfrac{1}{2}(\cos(\omega_s t - \omega_c t) + \cos(\omega_s t + \omega_c t))$$

After the filter, the signal becomes

$$\tfrac{1}{2}(\cos \omega_s t + \omega_c t)$$

The answer is C.

51. The Fourier series expansion for a pulse pattern is given by

$$f_2(t) = \frac{V_0 \tau}{T} + \frac{2V_0 \tau}{T} \sum_{n=1}^{\infty} \left(\frac{\sin \frac{n\pi\tau}{T}}{\frac{n\pi\tau}{T}}\right) \cos n\omega_0 t$$

The $n = 3$ element for the expansion is

$$\left(\frac{2V_0 \tau}{T}\right)\left(\frac{\sin \frac{3\pi\tau}{T}}{\frac{3\pi\tau}{T}}\right) \cos(3\omega_0 t)$$

Substituting $\omega_o = 2\pi/T$,

$$\left(\frac{2V_0 \tau}{T}\right)\left(\frac{\sin \frac{3\pi\tau}{T}}{\frac{3\pi\tau}{T}}\right) \cos \frac{(3)(2\pi t)}{T}$$

$$= \left(\frac{2V_0 \times 10^{-2} \text{ s}}{1 \text{ s}}\right)\left(\frac{\sin \frac{3\pi \times 10^{-2} \text{ s}}{1 \text{ s}}}{\frac{3\pi \times 10^{-2} \text{ s}}{1 \text{ s}}}\right)$$

$$\times \cos \frac{(3)(2\pi \times 10^{-2} \text{ s})}{1 \text{ s}}$$

[angles are in radians]

$$= 0.0196 V_0 \quad (0.02 V_0)$$

The answer is C.

52. The contact potential of a pn junction is given by

$$V_o = \left(\frac{kT}{q}\right) \ln \frac{N_a N_d}{n_i^2}$$

$$T = 50°C + 273.15°$$
$$= 323.15 \text{ K}$$

Solving for the acceptor concentration,

$$N_a = \left(\frac{n_i^2}{N_d}\right) e^{\left(\frac{V_o q}{kT}\right)}$$

$$= \left(\frac{\left(1.5 \times 10^{10} \frac{1}{\text{cm}^3}\right)^2}{7.0 \times 10^{17} \frac{1}{\text{cm}^3}}\right) e^{\left(\frac{(0.6 \text{ V})(1.602 \times 10^{-19} \text{ C})}{\left(1.38 \times 10^{-23} \frac{\text{J}}{\text{K}}\right) \times (323.15 \text{ K})}\right)}$$

$$= 7.4 \times 10^{11} \text{ cm}^{-3}$$

The answer is C.

53. Interleaving involves constructing memory from multiple modules where consecutive memory address locations are in different modules. When a request is made for data in sequential locations, the overall access time is faster than if all requests were processed by a single module.

The answer is B.

54. The control system circuit reduces to

$$\boxed{\frac{G_1}{1 \pm G_1 G_2}}$$

The problem statement specifies this is a negative feedback loop, so

$$\frac{C(s)}{R(s)} = \frac{s+3}{1+(s+3)^2}$$

The inverse Laplace transform is

$$\frac{C(t)}{R(t)} = e^{-3t} \cos t$$

The answer is B.

55. The poles and zeros of the root locus diagram for a control system are the poles and zeros of the open-loop transfer function. The open-loop transfer function is $G_C(s)G_1(s)G_2(s)H(s)$.

$$G_C(s) = s+3$$
$$G_1(s) = 1$$
$$L(s) = 0$$

$$G_s(s) = 1$$
$$H(s) = s+3$$

The open-loop transfer function is $(s+3)^2$.

There are no poles and two zeros at $(-3, j0)$.

The answer is C.

56. The input at time k is $x(k)$; the output at time k is $y(k)$; the input at time $k+1$ is $x(k+1)$; and so on. If the signal is DC, then $x(k) = x(k-1) = x(k+1) = x$ (a constant).

$$y(k) = \tfrac{1}{5}(2x + x + 2x) = x$$

$$\frac{y(k)}{x} = 1 \quad \text{[DC gain]}$$

The answer is A.

57. The impedance seen by the voltage source is the impedance of the capacitor plus the impedance of the parallel circuit.

$$Z = \left(-j3 + \frac{(5+j5)(-j5)}{5+j5-j5}\right) \Omega$$

$$= (-j3 + 5 - j5) \Omega$$

$$= (5 - j8) \Omega$$

$$= \sqrt{(5\,\Omega)^2 + (8\,\Omega)^2} \angle \tan^{-1}\left(\frac{-8\,\Omega}{5\,\Omega}\right)$$

$$= 9.43\,\Omega \angle -58° \quad (9.4\,\Omega \angle -58°)$$

The answer is C.

58. The currents into the node must sum to zero, so $i_2 = i_3 - i_1$.

$$i_3 = 1 \text{ A} \angle 58° = (\cos 58° + j \sin 58°) \text{ A}$$

$$i_1 = 0.707 \text{ A} \angle -116°$$
$$= (0.707 \cos(-116°) + j0.707 \sin(-116°)) \text{ A}$$

$$i_2 = i_3 - i_1$$
$$= (\cos 58° + j \sin 58° - 0.707 \cos(-116°)$$
$$\quad - j0.707 \sin(-116°)) \text{ A}$$
$$= (0.530 + j0.848 - 0.310 + j0635) \text{ A}$$
$$= (0.840 + j1.483) \text{ A}$$
$$= \sqrt{(0.840 \text{ A})^2 + (1.483 \text{ A})^2} \angle \tan^{-1}\left(\frac{1.483 \text{ A}}{0.840 \text{ A}}\right)$$
$$= 1.70 \text{ A} \angle 60.5° \quad (1.7 \text{ A} \angle 60°)$$

The answer is A.

59. The electric field sinusoidal wave equation for an isotropic homogeneous medium is

$$\nabla^2 E = -\omega^2 \mu_o \varepsilon E$$

This equation simplifies further since the electric field is only in the z-direction.

$$\frac{\partial^2 E_z}{\partial z^2} = -\omega^2 \mu_o \varepsilon E_z = -\omega^2 \mu_0 \varepsilon_r \varepsilon_o E_z$$

Substituting this electric field gives

$$-\tfrac{1}{3} \sin\left(10^8 t - \frac{z}{\sqrt{3}}\right)$$
$$= -\omega^2 \mu_o \varepsilon_r \varepsilon_o \sin\left(10^8 t - \frac{z}{\sqrt{3}}\right)$$

Solving for the relative permittivity and substituting equivalent units gives

$$\varepsilon_r = \frac{1}{(3)\omega^2 \mu_o \varepsilon_o}$$

$$= \frac{1 \, \frac{1}{\text{m}^2}}{\left((3)\left(10^8 \, \frac{\text{rad}}{\text{s}}\right)^2 \left(4\pi \times 10^{-7} \, \frac{\text{H}}{\text{m}}\right) \\ \times \left(8.85 \times 10^{-12} \, \frac{\text{F}}{\text{m}}\right) \right)}$$

$$= 3.00 \quad (3)$$

The answer is C.

60. The polarization is governed by the direction of the electric field vector. In this wave, the electric field vector is always in the z direction, so the wave is linearly polarized in the z direction.

The answer is A.

Trust PPI for Your FE Exam Review Materials

For more information, visit www.ppi2pass.com today.

Comprehensive Reference and Practice Materials

FE Review Manual
Michael R. Lindeburg, PE

- Complete FE exam coverage in 54 easy-to-read chapters
- Green text to identify equations, figures, and tables found in the *NCEES Handbook*
- Over 1,200 practice problems with step-by-step solutions
- 13 topic-specific diagnostic exams
- A 4-hour morning session exam
- A sample study schedule
- A comprehensive, easy-to-use index

Engineer-In-Training Reference Manual
Michael R. Lindeburg, PE

- A broad review of engineering fundamentals
- Over 980 practice problems
- More than 400 solved sample problems
- Over 2,000 equations and formulas
- Hundreds of tables and conversion formulas
- A detailed index

For complete solutions to all these practice problems, see the *Solutions Manual for the Engineer-in-Training Reference Manual*.

Discipline-Specific Review Series

6 Individual Books Cover Each of the Discipline-Specific Topics:

Chemical	Environmental
Civil	Industrial
Electrical	Mechanical

- Comprehensive review for each of the afternoon sessions
- 60 practice problems
- 2 complete 4-hour discipline-specific sample exams
- Complete step-by-step solutions

FE/EIT Sample Examinations
Michael R. Lindeburg, PE

- 2 complete sample exams
- 120 morning and 60 other disciplines afternoon session problems in each exam
- Complete step-by-step solutions
- Exam-like format, level of difficulty, time limit, and number of problems

1001 Solved Engineering Fundamentals Problems
Michael R. Lindeburg, PE

- Solve problems in the same multiple-choice format as the exam.
- Use the step-by-step solutions to understand how to efficiently reach the correct answer.
- Increase your problem-solving speed and confidence.
- Improve your ability to solve problems in SI units.

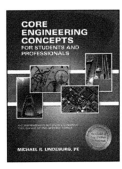

Core Engineering Concepts for Students and Professionals
Michael R. Lindeburg, PE

- Covers the breadth of an engineering degree
- Contains civil, mechanical, electrical, chemical, and industrial engineering subjects
- Features 82 chapters covering thousands of engineering concepts
- Contains more than 580 example problems with step-by-step solutions
- References over 3,700 essential engineering equations and formulas
- Presents Over 315 figures and 780 tables
- Lists fully-defined nomenclature for each chapter
- Includes a multiply cross-referenced index

Don't miss all the FE exam news, the latest exam advice, the exam FAQs, the unique community of the Exam Forum, Exam Cafe, and Passing Zone at **www.ppi2pass.com**.

Exam Cafe
Get Immediate Results with
Online FE Sample Exams and Practice Problems
Visit www.ppi2pass.com/examcafe.

PPI's Exam Cafe is an online collection of over 2,000 problems similar in format and level of difficulty to the ones found on the FE exam. Easily create realistic timed exams, or work through one problem at a time, going at your own pace. Since Exam Cafe is online, it is available to you 24 hours a day, 7 days a week, so you can practice for your exam anytime, anywhere.

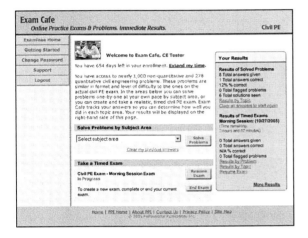

Solve over 2,000 FE Exam-Like Problems
- Problems are similar in format and level of difficulty to the problems found on the FE exam.
- Get complete coverage of the topics covered on the morning session of the FE exam as well as the other disciplines, chemical, civil, electrical, industrial, and mechanical afternoon sections.

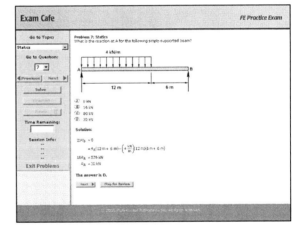

Create Realistic, Timed FE Exams
- Automatically create timed FE exams in seconds.
- Take exams that mirror the format, level of difficulty, and time constraints of the actual FE exam.
- Prepare yourself for the pressure of working under timed conditions.

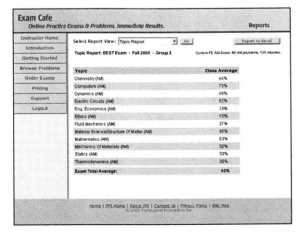

Immediate Online Solutions and Results
- Get immediate feedback with step-by-step solutions.
- View results of timed exams instantaneously.
- Analyze results by topic or by problem.
- Assess your strengths and weaknesses.

Visit Exam Cafe today at
www.ppi2pass.com/examcafe.